The Reluctant UXer's Field Guide to

Salesforce Design

Discover the joys and challenges you'll experience when you design for Salesforce.

Written by Stephy Hogan

Illustrated by Saifi Ullah

ISBN: 979-8-9902657-0-7

Library of Congress Catalog Number: 2024905862

Printed in the United States of America

First Printing: 2024

Cover design by Stephy Hogan & Saifi Ullah

Illlustrated by Saifi Ullah

Scan for companion
digital content.

Thank you to my amazing colleagues and friends at Red
Argyle, in general, and to those who lent me your eyes for
review: Michael Philbrick, Xavier Linsinski, Ashlyn Watters,
Nick Snyder, and .micah Eberman.

You've all helped make this complete and made sure I didn't
embarass myself by telling any of you incorrect information.

Table of Contents

Foreword

If you've ever created anything useful in the Salesforce ecosystem, then you recognize the unique challenges that it brings.

Those challenges come in the form of hard-fought success that sometimes looks pretty. More often it tends to look a bit dated and overly-sterile, though. (Just as the corporate software-heathens intended.)

That's because you haven't met Steph.

When we first got the chance to work together several years ago, it was obvious that she knew her shit. We first crossed paths during an interview round for a user experience role at a fairly successful self-proclaimed world-leading cybersecurity membership nonprofit called ISC² where we would wind up working tirelessly together to improve the user experience of our systems for our customers. Steph came highly recommended by a

mutual friend of ours who also gushed of her incredible talents. Within minutes of starting that interview I was in awe. I knew I needed her on our team and that our users would be better off with her as their champion.

Over the next few years, she helped disassemble the good-intentioned but poorly executed approaches to implementing Salesforce and web presences. From tearing down accessibility violations like floating action buttons (FABs), to rolling out simplified style guides, she got eyeballs-deep into the mess we created and helped us make it something to be proud of. She spent countless hours with us working through hundreds of LWC and Aura components designing a wonderful experience that looked beyond the confusing (and often conflicting) vernacular and concepts that Salesforce has given us.

After our paths took their own natural routes and we left ISC² to pursue our own adventures elsewhere, I've regularly sought her out to help on projects I've been involved in and to recommend her to others that could benefit from her unique ability to turn a cold and drab user interface into an intuitive step along a delightful user journey.

Steph continues to cut through the bullshit and give us what we all want: a break from the chatter that lets us focus on delivering pretty & usable apps—one CPE Portal and floating accessibility overlay at a time.

Nick Snyder

Lecturer @ RIT, EMBA, B.S., AAGG, CTO of Higher Booking, Pontificator

Illustrated by Stephy, not Nick.

Note from author: Nick gave me that title as a joke. I went with it. You're welcome, friend.

3

Introduction

"Space is big. You just won't believe how vastly, hugely, mind-bogglingly big it is. I mean, you may think it's a long way down the road to the chemist's, but that's just peanuts to space."

—Douglas Adams, The Hitchhiker's Guide to the Galaxy

Replace the instances of "space" with "Salesforce" in the quote above and you get the phrase that runs through my head whenever someone asks what I work on for a living.

Seriously. There's so much to this global beast of a CRM that just when you think you're beginning to get your head wrapped around it, they announce a whole new, game-changing feature-set along with documentation, videos, Trailheads, a kick-youin-the-nethers-level-of-difficulty Certification to study for, and a brand new cute and fuzzy mascot that must be obtained in at least six outfit variations.

While you may be detecting notes of good ol' Gen X

resentment and malaise, it's actually not the case this time. I mean, Salesforce really IS big... powerful...and it sure seems like you can find a way for the system to do just about anything a business could want.

And there lies the challenge. What do you do with the system that can do anything?

Where do I start?

What should I use?

What should I NOT use?

Do I turn on all of these cool components and features just because I can?

HECK YEAH.

So. Many. Checkboxes.

So. Very. Shiny.

And for a number of years, that was the idea. When in doubt, turn on the cool new thing and see what happens.

Now, I'm not saying that newly-minted Salesforce Admins ran amok like monkeys in a banana factory, but it has been observed that when you have a very powerful, feature rich system...the kind that is spendy enough that you should really try to get your money's worth out of it... that the tail can sometimes wag the dog and the needs, goals, and desires of the target audience have all become secondary to what the system can do.

So you get a feature-centric product launch. Your new product launches with a complex interface that includes high-tech dashboards, multiple toolsets, and extensive configuration options. The marketing materials focus on the sophisticated technology and the breadth of functionalities, promoting the product as a cutting-edge solution.

After launch, users start to interact with the product, and issues begin to surface...

- **Complexity and Usability Issues:** Users find the interface overwhelming and difficult to navigate. The multitude of features, while technically impressive, does not align well with the everyday needs of the user.
- **Lack of Relevance:** The features, though advanced, often don't address the primary challenges or goals of the target audience. Users struggle to see how the tool fits into their daily workflows.
- **Poor Adoption and Frustration:** Feedback reveals that users are frustrated with the steep learning curve and the perceived lack of practical utility. Adoption rates are lower than expected, and customer satisfaction scores are disappointing.

So what's the business impact here? The focus on technology rather than user needs results in a product that fails to meet business goals like user retention, sales growth, and positive brand association. The company is forced to reconsider its approach and potentially pivot towards a more user-centered design.

It's time for the weirdo dressed in black.

RELEASE THE KRAKEN...er... "SALESFORCE DESIGNER".

Now, this is the part where I'd settle in under my covers...just a little tyke clutching my Astro plushy and waiting for my father to read the "happily ever after" where our hero, the Salesforce Designer-slash-Strategist, turns it all around.

Realizing the mismatch, the company begins to collect user feedback, undertakes rounds of usability testing, and the team works on aligning the product's capabilities with the actual needs and goals of their target audience. Just like in the days of yore!

Simplify the interface!

Prioritize key features based on user demand!

Redefine the product's value proposition to better meet market expectations!

Except here's the problem, friends.

You know that "Salesforce Designer" I told you about? They're the stuff of myths and legends. They're what Unicorns talk about when they need an example of something rare and unbelievable.

Wild UX Architects and UI Innovators may be plentiful in the broader digital world of which you may be more familiar, but in the Salesforce Ohana, whispers of their existence dance along SOMETHING.

For a number of years, this has been the official Salesforce Design methodology...

1. Find a designer. Any area of expertise will do.

2. Hand them the largest can of Monster energy drink you can find; wish them good luck

3. Let them go ham on a design from scratch, as long as it looks like Lightning

4. Custom code the components and interface they came up with, going over budget and creating a system-update nightmare

5. Depending on the levels of frustration, either go back to #3 or #1...rinse...repeat

Finally, it was time to stop the madness. We needed real, certified, trained, and knowledgeable Salesforce Designers. And thus Salesforce #DreamDesigners came to be.

In recent years we've seen more and more advancements in the training and evolution of the much-needed and very desirable art of Salesforce Strategy and Design. Teams are recognizing the value that the disciplines bring to the table...software and sites that are developed via Design Thinking and Human-centered Design, and thus more successful in meeting user needs and business goals alike.

Designs that leverage the power of OOTB objects, native components, and thoughtful means of extending functionality via AppExchange.

There are now more design-centric Salesforce Trailheads, Certifications, and prototyping resources than ever before.

And the demand for talented designers that can get the most out of the Salesforce Customer 360 ecosystem continues to grow in lockstep with Enterprise undertaking meaningful digital transformation efforts with wide-ranging, global impact.

Salesforce Designers are MADE, not born.

That's what excites me about this here book that Stephy and her pals have created just for you. Hopefully you'll come out the other side of this learning experience with a smile on your face, some more booklearnin' in yer head, a new skip in your step, and you're ready to be the next designer to take this discipline and practice to the next level.

Because really, Salesforce is big. You just won't believe how vastly, hugely, mindbogglingly big it is.

And you can use it to solve big problems and make great things.

Alright, have at it. Grab a can of Monster, and good luck.

.micahEberman

UX Architect, Lead Strategist, and Salesforce #DreamDesigner

Illustrated by .micah

Am I really designing for Salesforce?

You've either acquired this guide for yourself or some genius of a human being gifted it to you. So, yes. You'll be designing for and IN Salesforce.

Have you ever worked on sites powered by a content management system? Just think of Salesforce as a beefed up version of a very custom WordPress site—where the terminology is completely different. Instead of a content management system (CMS), Salesforce is a customer relationship management system (CRM).

It holds a veritable @&#$?&!-ton of data. Companies in all industries use it not just for relationship management. Salesforce is also used to record continuing education credits, for hotel reward program members to book their next vacation, and to track the efficiency of your bulk Doritos shipment.

The design possibilities are endless. So if you find yourself in the world of Salesforce design and feel overwhelmed, this field guide is for you. It's the guide I wish existed during my Trial by Fire entry into Salesforce.

First things first— let's talk jargon.

I'll take "Things every designer new to Salesforce will say at some point" for $100.

If you take how a simple spreadsheet is structured, you'll find that there is still no overlapping vocabulary between it and Salesforce.

You know what a column is.

You know what a row is.

You know what a cell is.

Same goes for fields, right?

Now throw all of that knowledge out the window while we run through how Salesforce names things. Trust me on this one. Learn it now and save a lot of confused head scratching later.

13

Shared language? There is none.

I kid, I kid. Sort of. Not really. If there's one thing companies like to do is create their own way to say things. Because it's cool.

OBJECTS Going back to the spreadsheet reference, there is a **tab** at the bottom with a name: Contacts. The name of the tab is the equivalent to an Object in Salesforce.

RECORDS Records are the rows of the spreadsheet. That should be easy to remember, right? R and R? I like it. Let's go with that. In this example, each contact and their corresponding info is all in one row. So "contact" is the type of record.

FIELDS And last but not least... Fields. I know. The first thing you think of is a form field, a way to collect data. In the spreadsheet, the **columns** are the fields, or more specifically, the field headers.

Which, at least, makes sense because we use form fields to gather that kind of information.

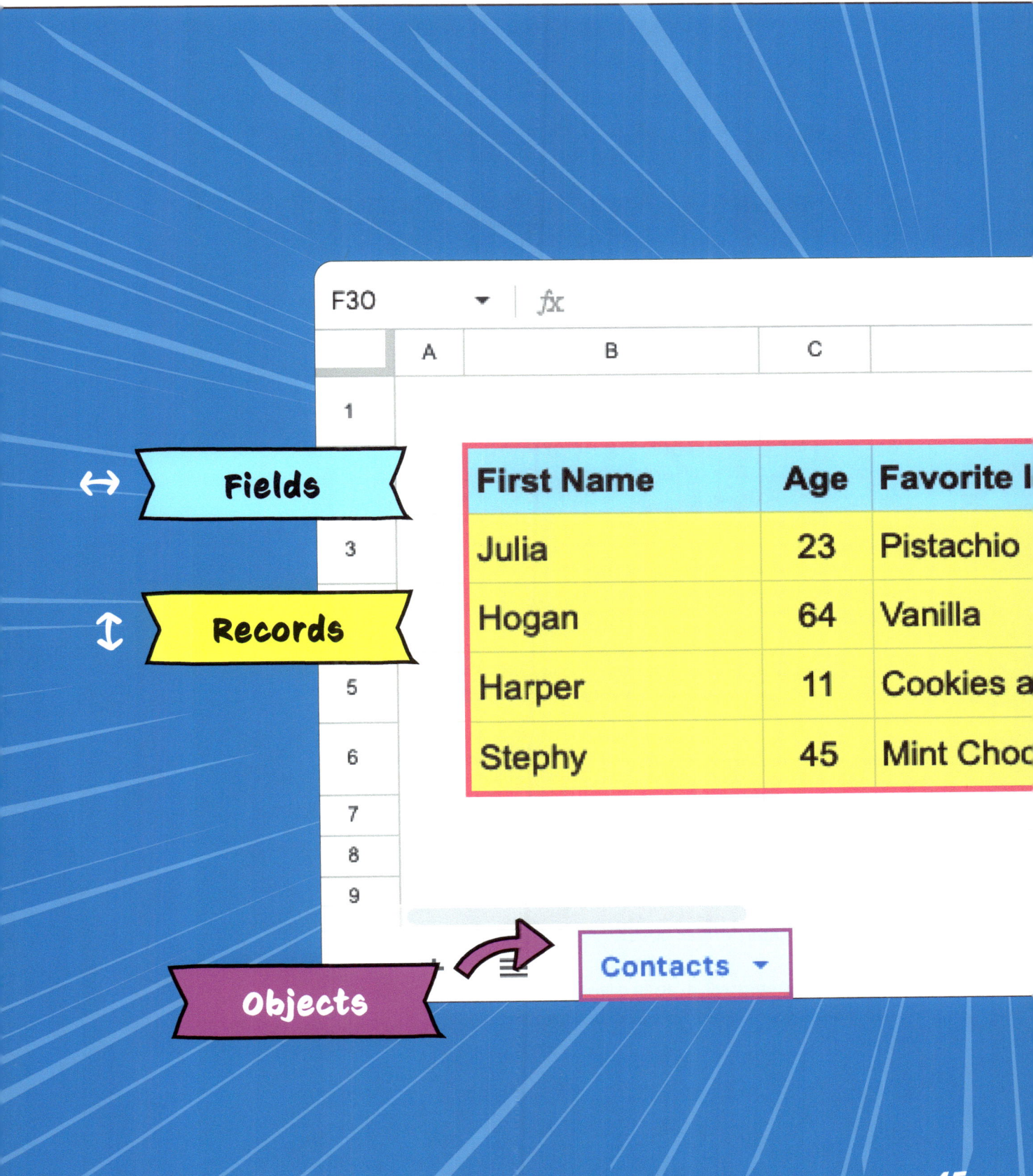

F3O   fx

	A	B	C	
1				
		First Name	**Age**	**Favorite I**
3		Julia	23	Pistachio
		Hogan	64	Vanilla
5		Harper	11	Cookies a
6		Stephy	45	Mint Choc
7				
8				
9				

Fields

Records

Objects

Contacts ▾

There's "classic" and "lightning" versions.

Usually "classic" refers to something that has a timeless elegance that never goes out of style. Not here.

You'll hear the word "lightning" thrown around a LOT. It can mean VASTLY different things depending on the person who is using it.

People who work with Salesforce on a somewhat regular basis usually mean the Lightning version and not the antiquated Classic view. (I mean LOOK at that. It screams late '90s.)

Developers are referring to the same thing but they usually are using it as the counterpart to an Experience Cloud project (we'll get to that soon.)

And sometimes they'll use it to refer to what kind of custom component is being built (as opposed to "aura." And yes, we'll touch on all of that soon, too.)

If a designer uses the word "lightning," depending on how long they've designed for Salesforce, it could refer to:

• Lightning Experience, or LEX (vs. Classic)
• Lightning Web Components, or
• Lightning Design System (usually this, but not always... And we'll talk about this in a hot second).

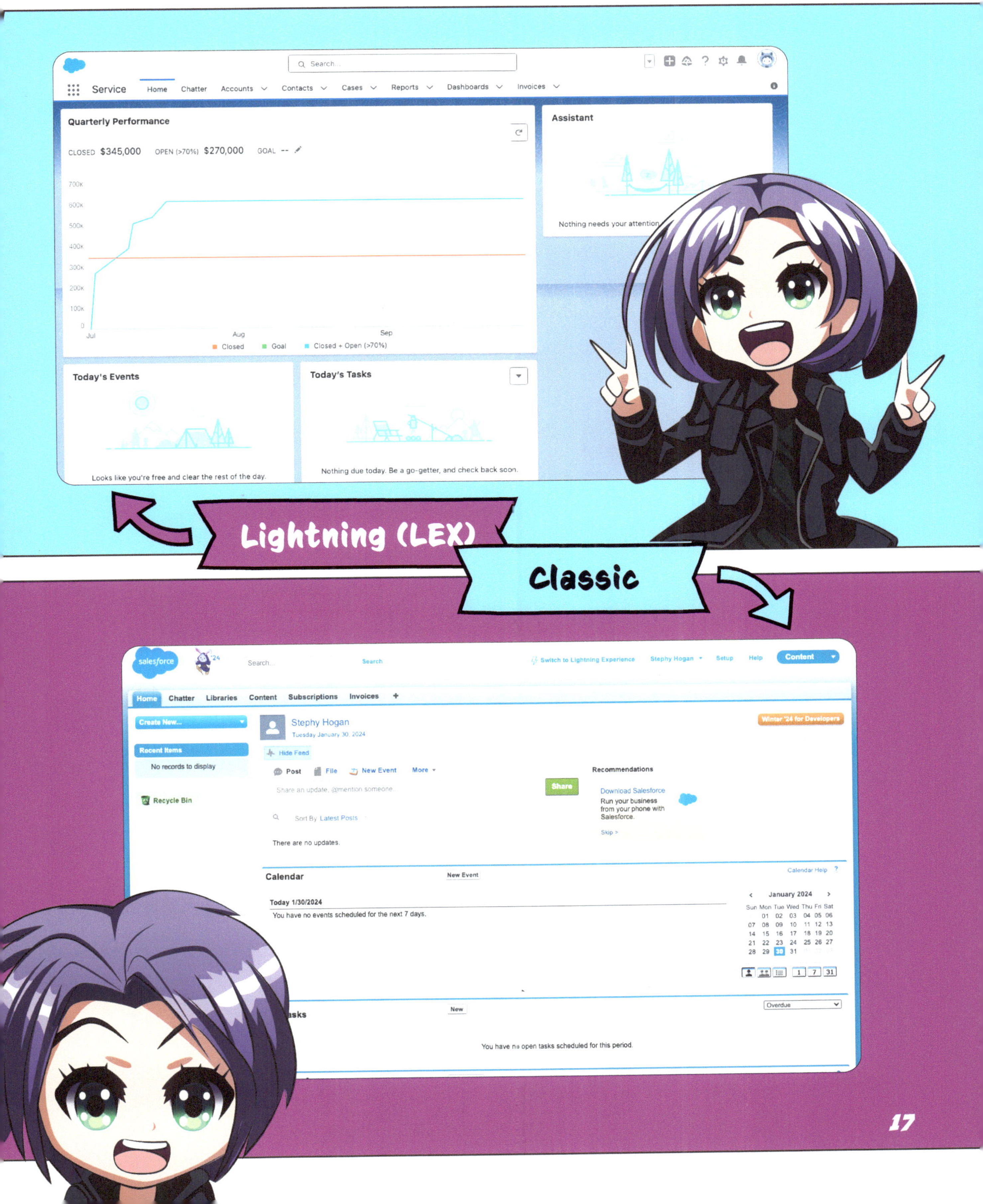

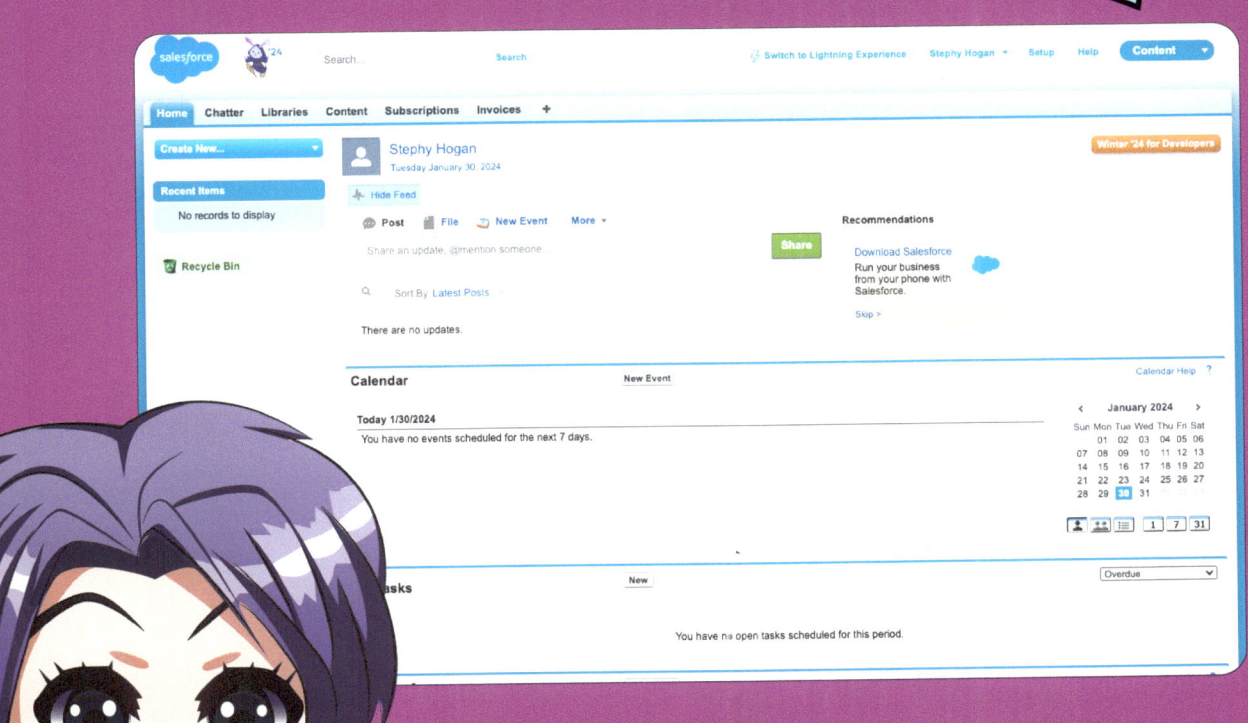

Lightning (LEX)

Classic

Lightning apps are not your traditional apps.

"An app? That's not a..." Remember when I said that? It was only 6 pages ago. Sooooo, yeah...these aren't like any app you're used to.

When you're sitting there, looking at Salesforce Lightning (and possibly questioning your life choices), you'll see:

- A waffle (yum)
- A couple of words (in this example "Service")
- Tabs across the top

This is an app. Tabs are objects in that app (though not all of them).

How do you change apps? You use the waffle menu. When you select the waffle, Salesforce will show you 7 more apps you can choose from.

Don't let that fool you. When you choose "view all," you'll get a small glimpse at how incredibly beefy Salesforce can be. But why do you care?

The great thing about apps is that you can create apps that hold personalized processes, dashboards, and other insights based on individual roles (sales, support, educator, student, etc.).

Anyone hungry for Waffle House now? I am.

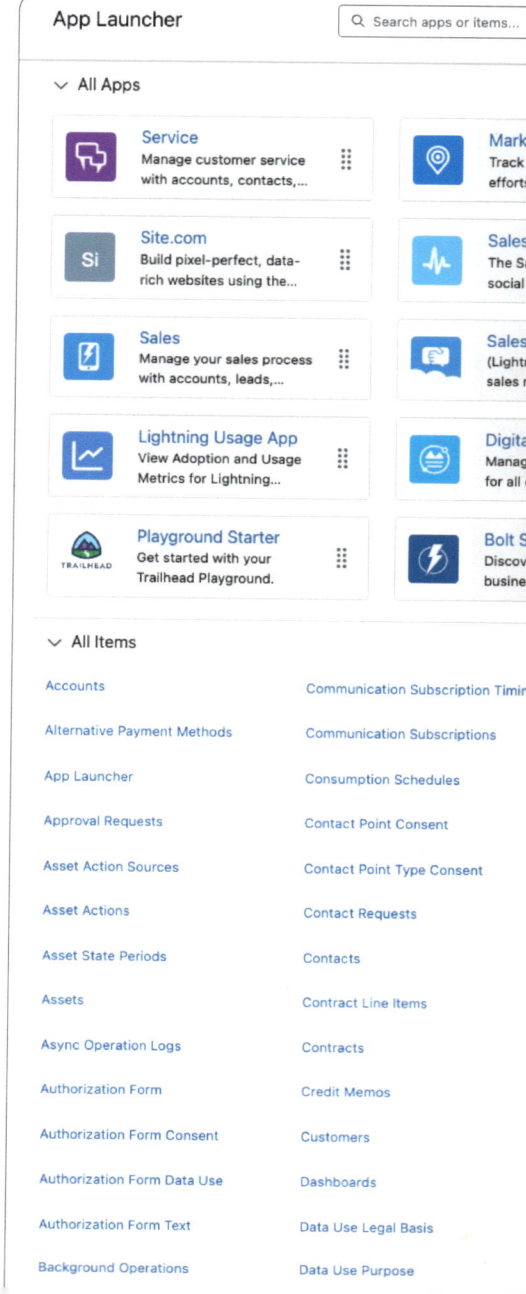

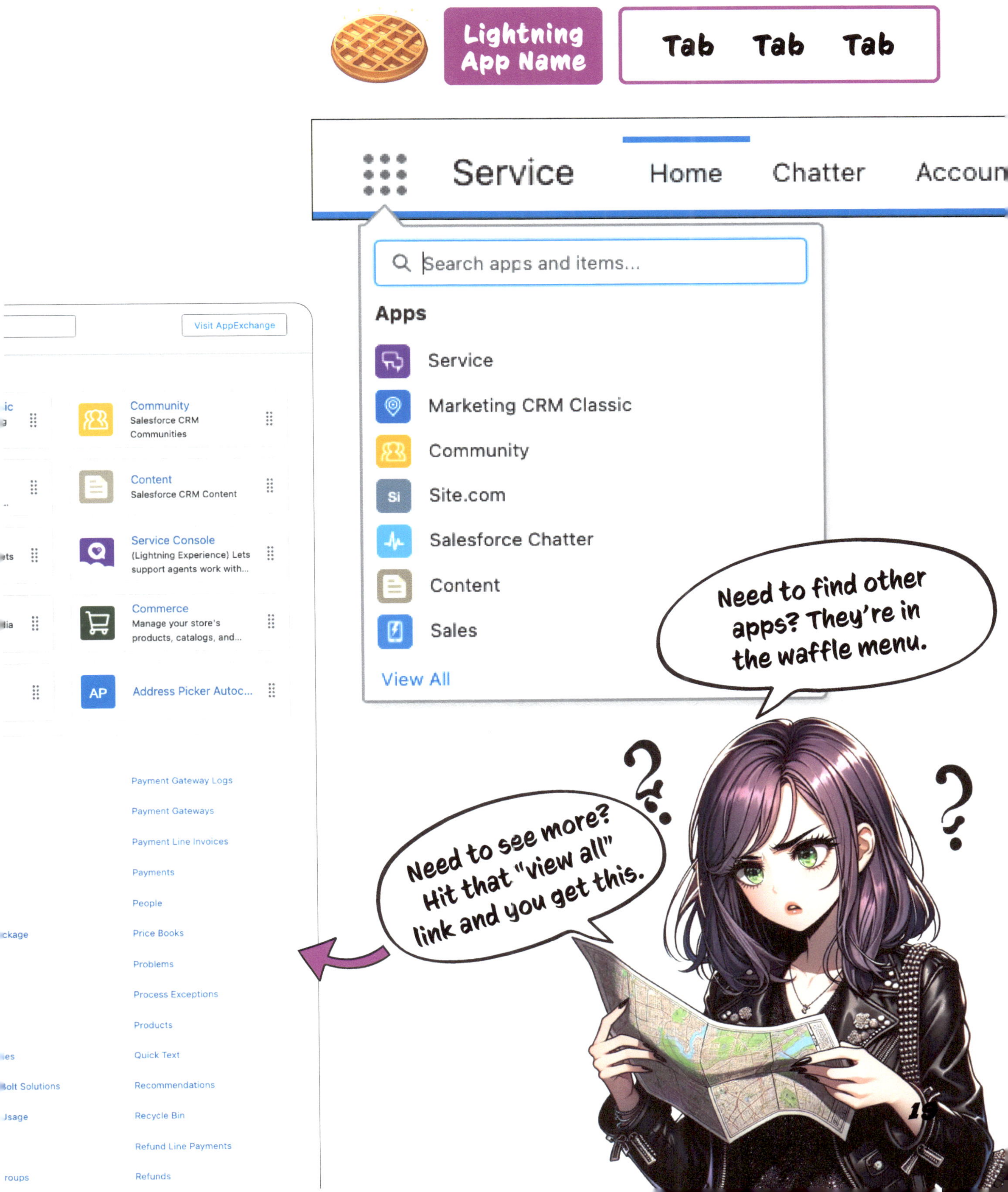

Lightning App Name

Tab Tab Tab

::: Service | Home | Chatter | Accoun

Q Search apps and items...

Apps

- Service
- Marketing CRM Classic
- Community
- Site.com
- Salesforce Chatter
- Content
- Sales

View All

Visit AppExchange

Community
Salesforce CRM Communities

Content
Salesforce CRM Content

Service Console
(Lightning Experience) Lets support agents work with...

Commerce
Manage your store's products, catalogs, and...

AP Address Picker Autoc...

Payment Gateway Logs

Payment Gateways

Payment Line Invoices

Payments

People

Price Books

Problems

Process Exceptions

Products

Quick Text

Recommendations

Recycle Bin

Refund Line Payments

Refunds

Need to find other apps? They're in the waffle menu.

Need to see more? Hit that "view all" link and you get this.

Community vs. Communities vs. Experience Site

Those 3 things actually refer to only 2 things. And you can tell who's been in Salesforce longer by the language they use.

(TRAILBLAZER) COMMUNITY

The Trailblazer community hangs out on Trailhead and is a great place to go to learn how Salesforce works and how to do things in it. They have interactive learning paths called Trails that you can earn badges for. It's a decent resource.

EXPERIENCE SITE

Experience sites are where designers thrive in the Salesforce environment. They're basically websites with Salesforce as the database. We'll get into these later.

When the Spring '21 release was pushed out, designers and developers suddenly saw "Digital Experiences" in place of "Communities."

COMMUNITY

Same thing as an Experience Site, just an older name. **All the cool kids call them communities.** (It's me, I'm the cool kid. Take that for what it's worth.)

Community

Communities

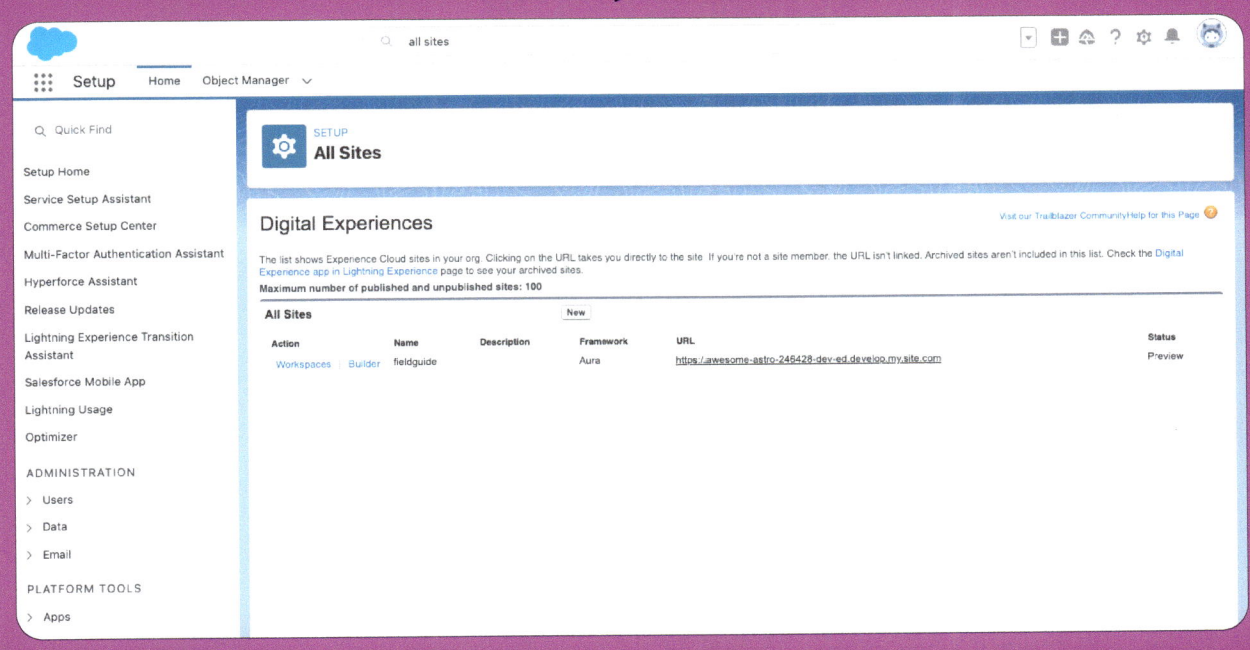

Salesforce Clouds

There are so many clouds in the Salesforce sky that the weather would be perpetually overcast. All in all, there are 15 clouds (!!!), but we'll focus on the 6 main ones.

SALES CLOUD

Tracks customers and the organizations they're part of. Tracks their partners and competitors. Track the life cycle of a sale. Loads of automation.

> **Design stuff like:**
> Efficient Sales Journeys • Data viz

SERVICE CLOUD

Helps call centers and customer service folks to track customer issues and their resolutions.

> **Design stuff like:**
> Help Ticket Processes • Self-service Portals

MARKETING CLOUD

A whole bunch of offerings smushed into one cloud. Email campaigns, marketing automation, lead management, blah blah blah.

> **Design stuff like:**
> Email templates • Landing pages • Data viz

COMMERCE CLOUD

This is the one that turns Salesforce into something that can build and manage (the back end of) a storefront.

> **Design stuff like:**
> eCommerce user journeys • Whatever the opposite of storefronts is

ANALYTICS CLOUD

This one is still co-branded with Tableau's logo (at least at the time of this being published.) Straight from the horse's mouth: Spot opportunities, predict outcomes, get recommendations, and more with CRM Analytics.

> **Design stuff like:**
> Data viz dashboards • Other stuff

EXPERIENCE CLOUD

This is the best one. I might be biased. With this you can make self-service portals, partner portals and all kinds of websites that look like websites and not like Salesforce at all. Designers have the most fun here. All Experience sites are powered by the other clouds.

> **Design stuff like:**
> eCommerce sites, Support sites, B2B2C portals, ANYTHING!

I'm smiling on the outside, but on the inside I'm twitching.

They update the platform often.

Salesforce pushes out updates quarterly. And sometimes there are some pretty exciting design updates. (By "exciting" I mean "coming to the 21st century," but they still bring me joy.)

You can find a link to the release notes at the bottom of help.salesforce.com. It will take you to the most recent set of notes, but you can always change which release you're referencing.

The left column gives you the overview of the veritable MILLIONS of updates in the release. Don't try to look through that list for anything related to design. It's not there. Yet.

I highly recommend that you just use the search field for "experience," "design," and "accessibility." You'll always find at least ONE update that pertains to what we do.

Usually.

Help

Salesforce Sp

CONTENT

Q Search

▼ Filter by (0)

∨ Salesforce S
   What's Ne
   Notes?
   How to U
   Get Ready
› Monthly F
   Release N
   How and
   Available?
› Supportec
› Salesforce
   Commerc
› Customiza
› Data Clou
› Deployme
› Developm
› Einstein

help.salesforce.com > release updates

Product Support ∨   My Cases   **Contact Support**

se Notes   Spring '24   ▼   PDF ⬇

← [back arrow button]

+ Add

Notes

rce Release

tes

s Become

SALESFORCE HELP > DOCS > SALESFORCE RELEASE NOTES

Salesforce Spring '24 Release Notes

See how the Spring '24 release helps teams work smarter with new product innovations built on Data + AI + CRM + Trust.

- **What's New for the Salesforce Release Notes?**
  Learn about new features that make the Salesforce release notes easier to use. Think of this page as the release notes for the release notes and check back each seasonal release to see what's new and improved. We also welcome your feedback.

- **How to Use the Release Notes**
  Our release notes offer brief, high-level descriptions of enhancements and new features. We include setup information, tips to help you get started, and best practices to ensure your continued success.

- **Get Ready for the Release**
  Reading the release notes is a great step in preparing for the release. These other resources help get you and your users ready for what's coming your way. We add resources throughout the release when they become available, so check back often.

- **Release Notes for Features Released Monthly**
  Salesforce releases features and enhancements more frequently than three times per year for some products. Find out what's new and read more about these features, as often as monthly, right here in the seasonal release notes.

- **Release Note Changes**
  Read about changes to the release notes, with the most recent changes first.

- **How and When Do Features Become Available?**
  Some features in Spring '24 affect all users immediately after the release goes live. Consider communicating these changes to your users beforehand so that they're prepared. Other features require direct action by an administrator before users can benefit from the new functionality.

- **Supported Browsers**
  We've made some changes to our supported browsers documentation, making it easier to find what you need. Supported browsers for Salesforce vary depending on whether you use Salesforce Classic or Lightning Experience.

- **Salesforce Overall**

SALESFORCE

Privacy Statement

Security Statement

Terms of Use

Participation Guidelines

Cookie Preference Center

☑✗ Your Privacy Choices

COMMUNITY

AppExchange

Salesforce Admins

Salesforce Developers

Trailhead

Training

Trust

SUPPORT & SERVICES

Need help? Find more resources or connect with an expert.

Get Support

Powered by Experience Cloud

English   ▼

6389

[social media icons]

Not all users get the same features.

If Salesforce versions and different Salesforce clouds haven't confused you enough...

Salesforce bundles feature sets together and calls them editions. They otherwise look the same. While anything is possible with Salesforce, the level of effort and cost to build your gorgeously designed experiences will vary greatly depending on which edition has been purchased.

Starter and Pro Suite are designed for smaller companies starting out and offer a more well-rounded set of features. They're the newest versions. Of course, they have limitations (you can't customize ANYTHING in starter, for example).

In Professional, you CAN customize anything.

In Enterprise, you get to do super advanced workflows and other salesy-type things. Devs can build APIs to connect with non-Salesforce systems. And you get unlimited automations, user roles, page layouts, etc.

In Unlimited, you get super special top-notch extra support from Salesforce should you need it.

In case you wanna know the details...

	Starter	Pro Suite	Professional	Enterprise	Unlimited
Sales					
Account, Contact, Lead, and Opportunity Management	✓	✓	✓	✓	✓
Email Integration	✓	✓	✓	✓	✓
Task Management and Activity Feed	✓	✓	✓	✓	✓
Customizable Reports and Dashboards	✓	✓	✓	✓	✓
Salesforce Meetings	✓	✓	✓	✓	✓
Direct Payment Links for Opportunities	✓	✓	✓	✓	✓
Payment Portal for Payments	✓	✓	✓	✓	✓
Forecast Management		✓	✓	✓	✓
Quote Creation and Management		✓	✓	✓	✓
Direct Payment Links for Quotes		✓	✓	✓	✓
Service					
Integrated Email Support	✓	✓	✓	✓	✓
Case Management	✓	✓	✓	✓	✓
Knowledge Management	✓	✓	✓	✓	✓
Custom Email Templates	✓	✓	✓	✓	✓
Macros		✓	✓	✓	✓
Omni-Channel Routing		✓	✓	✓	✓
In-App and Web Messaging		✓	✓	✓	✓
Marketing					
Email Campaigns	✓	✓	✓	✓	✓
Smart Segmentation	✓	✓	✓	✓	✓
Premade Email Templates	✓	✓	✓	✓	✓
Drag-and-Drop Email Builder	✓	✓	✓	✓	✓
Einstein Send Time Optimization	✓	✓	✓	✓	✓
Campaign Analytics	✓	✓	✓	✓	✓
Enhanced Capabilities					
Drag-and-Drop Email Builder	✓	✓	✓	✓	✓
Einstein Send Time Optimization	✓	✓	✓	✓	✓
Campaign Analytics	✓	✓	✓	✓	✓
Custom Apps		✓	✓	✓	✓
Custom Objects		✓	✓	✓	✓
Process Automation with Flows		✓	✓	✓	✓
Approval Processes		✓	✓	✓	✓
Sandbox for Testing and Training		✓	✓	✓	✓
AppExchange Access		✓	✓	✓	✓

Doing ok so far?

Doodle space.

Doodle space.

Good. Have a
gummy bear.

30

Ok, how do I make it look less... Salesforcey?

Clients ask me this ALL. THE. TIME. There will be times you will get to customize the @*&% out of the interface. And other times? You'll be stuck with it. That doesn't mean you can't make Salesforce FEEL less (stereotypically) Salesforcey.

UX isn't always part of a Salesforce solution. It should be.

Most Salesforce build teams I've come across don't include UX unless they're making an Experience Cloud site or a custom component for Lightning. That needs to change.

Designer mindset needs to change as well. That's why you're here. (I mean, it might not be the primary reason, but I'm hoping you'll join me on the dark side.)

Experience Cloud is where many (dare I say MOST) clients and engineers think UX belongs. And ONLY there. While it's essential that UX is the star of the show when building those projects, we also need to be involved in Lightning (I usually refer to that as the back end because I started in Experience Cloud. Forgive me if I start using that term at any point.)

What can we do in new org setups, org cleanups, org splits, and other non-Experience Cloud projects?

We apply the foundational elements of good user experiences. Keep things simple. Don't overwhelm the user. Make it intuitive. Make it "just work."

Create a Salesforce environment that makes users more productive, efficient, and enjoyable.

INCREASED USER ADOPTION

One of the biggest issues that companies have with their Salesforce orgs is low user adoption. I'm not surprised. By now, you've *seen* the backend of Salesforce at least once. You know how overwhelming it can be. We can fix that with UX design.

INCREASED CLARITY

Information overload describes Salesforce Lightning rather well, don't you think? Look, once we get our grubby little design hands on it and work our magic, suddenly confusion about what the user needs to do is now immediately obvious. The information they need to know is ridiculously easy to find and understand.

DECREASED TIME TO TASK COMPLETION

Doing things in Salesforce means making new records, each in their own section. Then, to make sure they're connected properly, you have to go back and manually connect them. (Not always, but more than anyone wants to.) If we shave a couple of seconds off of repeated tasks, we can save HOURS of time per week.

DECREASED NEED FOR SUPPORT

And of course, if a system is intuitively easy to use, and its information is discoverable and under-standable... users will inevitably need less help using it.

Custom vs. Out-of-the-box (OOB) or (OOTB)

Until a few years ago (maybe more, time flies), if you wanted to design usable user journeys in any part of the Salesforce ecosystem, custom components had to be built.

I'm going to just start with a data table as an example. We call them List Views in Salesforce and they can be SUCH a pain in the butt. With OOB tables, you don't really have any way to add more visual elements or control padding.

Yes, some mild front-end development will get you most of the way there, visually-speaking. When it comes to functionality, in the custom table on the bottom of page 35, the filters, the way the table prints, and the pagination are all custom code.

The out of the box table (top of page 35) takes about **5 minutes** to put on the screen—maybe 20 minutes overall to build, test, and get approved.

The custom table took **4 months** to develop (because don't forget about stakeholder approvals, iterations, random code bugs, and QA!).

Takes about 20 minutes and zero code.

Name	Type	Start	End	Credits	Group	On Your Behalf	Status
InfoSec: Content & Intro to a long name	Education	05/05/2018	05/05/2018	8	B	No	Audit
InfoSec: Content & Intro to a long name	Education	05/05/2018	05/05/2018	8	B	Yes	Accepted
InfoSec: Content & Intro to a long name	Education	05/05/2018	05/05/2018	8	B	No	Accepted
InfoSec: Content & Intro to a long name	Education	05/05/2018	05/05/2018	8	B	No	Accepted
InfoSec: Content & Intro to a long name	Education	05/05/2018	05/05/2018	8	B	Yes	Accepted

Out-of-the-box

Custom

Takes about 4 months and custom code.

 **100.00 CPEs Completed of 120**
Last updated for CISSP on 06/10/2018

Current Year   Current Cycle   Previous Cycle   🖨 print pdf

Name ▲	Type ▼	Start ▼	End ▼	Credits ▼	Group ▼	On your behalf ▼	Status ▼
InfoSec: Content & Intro to a long name	Education	05/05/2018	05/18/2018	8	B		AUDIT
InfoSec: Content & Intro to a long name	Education	05/05/2018	05/18/2018	8	B	✓	Accepted
InfoSec: Content & Intro to a long name	Education	05/05/2018	05/18/2018	8	B		Accepted
InfoSec: Content & Intro to a long name	Education	05/05/2018	05/18/2018	8	B		Accepted
InfoSec: Content & Intro to a long name	Education	05/05/2018	05/18/2018	8	B	✓	Accepted

Don't see your CPE? Contact Member Support for help. ❯

Items Per Page
5   ⌄

Back

You can totally brand your Salesforce org

A little bit anyway. Is it complex?
Not very. They're called Themes.

You can set up a default brand theme or themes in Salesforce Lightning setup.

- Brand Image (600 by 120px)
- Primary brand color
- Page background color
- Global header background color
- Link color (primary color or standard Salesforce blue)
- Page background image (1800 x 360px)
- Default group banner image (1800 x 360px)
- Default user profile banner (1800 x 360px)
- Default group avatar (200 x 200px)
- Default user profile avatar (200 x 200px)

The neat part is that you can have branding sets and show a different org style to different types of people. Sales could see one skin, Service could see a different one. Or if you're users are from, let's say, two different companies, you can create and show each company a different theme. Personalization is fun.

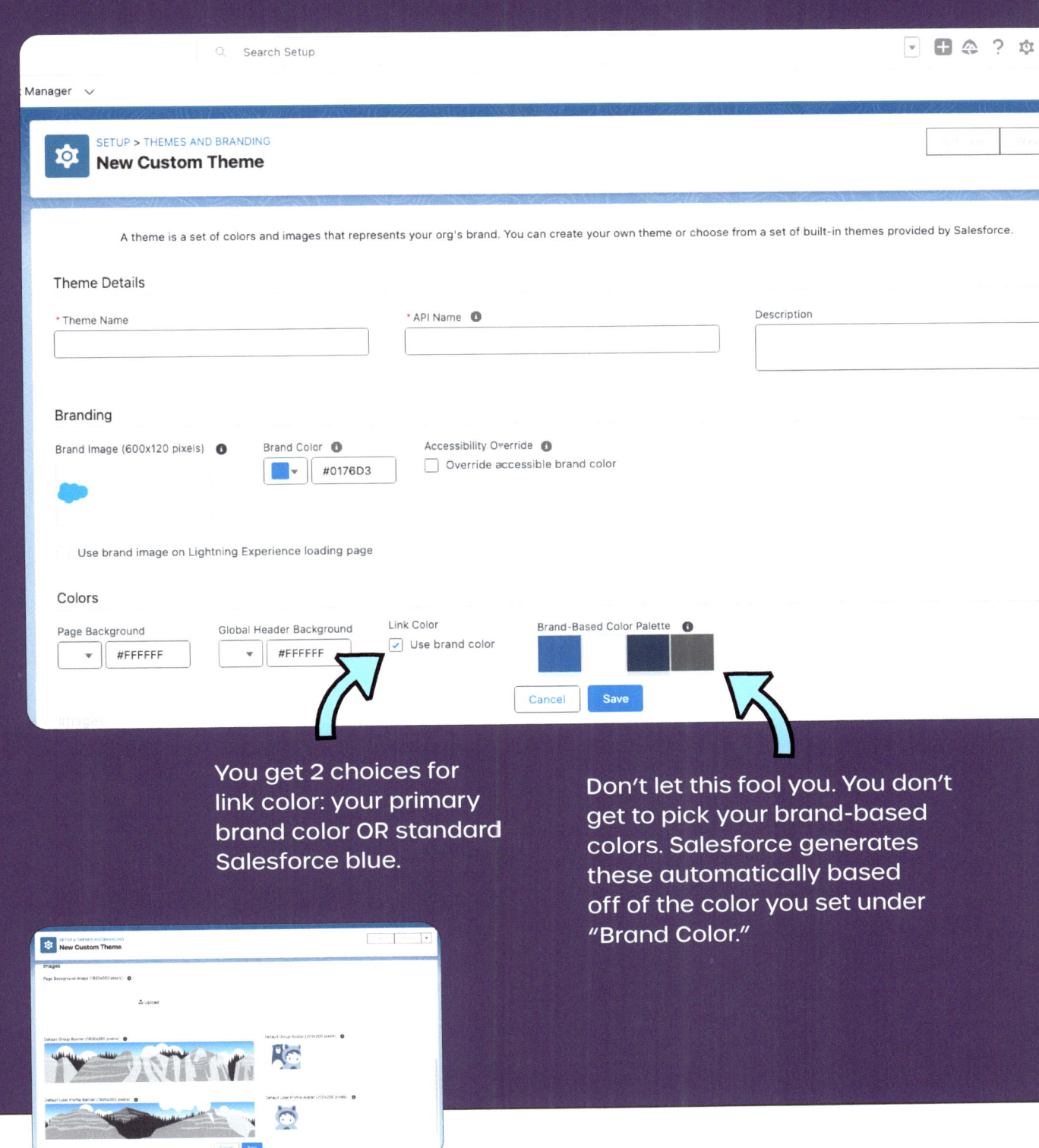

You get 2 choices for link color: your primary brand color OR standard Salesforce blue.

Don't let this fool you. You don't get to pick your brand-based colors. Salesforce generates these automatically based off of the color you set under "Brand Color."

The bottom half of the page is where you set images that aren't the "brand image."

Page layouts are dead! Long live page layouts!

If Salesforce had to choose a descriptor for its relationship with page layouts, it would be, "It's complicated."

Have you ever built something in the latest version of Adobe InDesign and then handed it off to a client only to discover they can't open it because they have an old version of InDesign? I have. The lack of backwards compatibility always annoyed the s$&% out of me. PowerPoint was guilty of it too.

Well. It seems Salesforce "solved" that problem by building the new version around the old version. Lightning is the industry standard but some companies still use Classic, and Salesforce lets them.

Page layouts are a relic of Salesforce Classic. Could Salesforce phase them out eventually and only have Lightning pages? I like to think it's possible—even if it would require a minor revolution for both Salesforce customers and Salesforce themselves.

Lightning pages are where Salesforce has a chance to shed its stereotypes of being bulky, overwhelming, and ugly. (Sorry, Salesforce, I have a love/hate relationship with you.)

Companies don't usually think about having UX designers help implement Lightning pages even though user experience is just as important at this level of a build as it is in Experience Cloud sites.

If UX pairs up with engineering, you can help companies push through the perception that users either have to muddle through a messy org every day or spend millions of dollars to start over. Because they don't have to do either of those things. Designers create the journey, developers build it out on a Lightning page using dynamic forms and screen flows. Declarative UX. Magic.

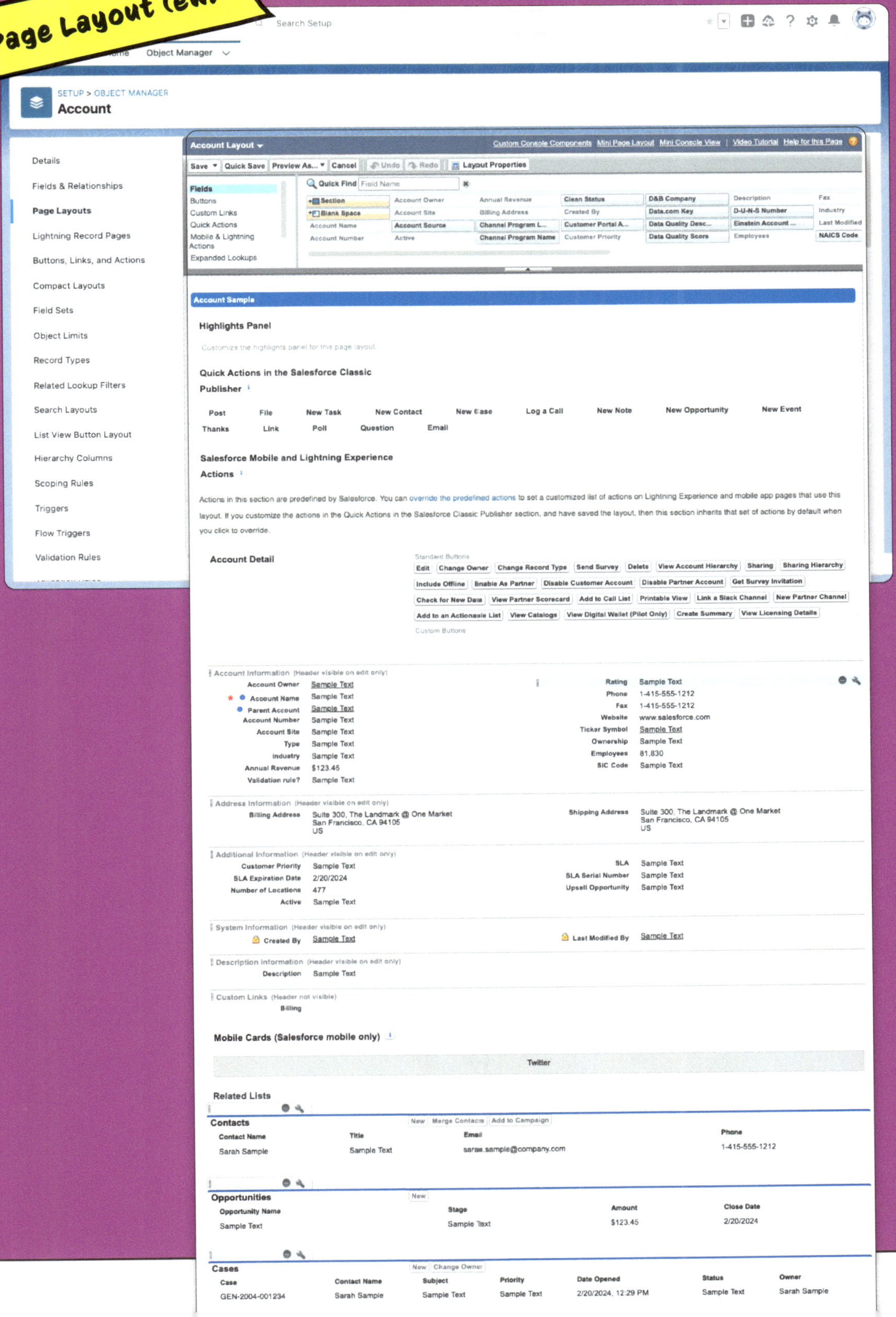

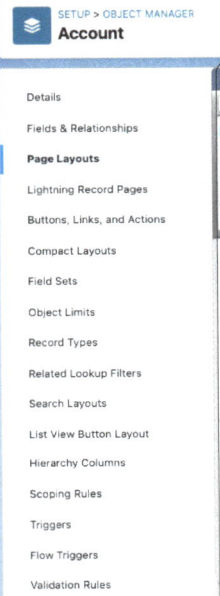

Let's talk about best practices.

Maximize usage of Lightning pages, screen flows, and dynamic forms.
Once you do, you start thinking "how do I want this to work" instead of "what will Salesforce let me do."

Create one lightning page per broad role.
(i.e. don't go too granular, you don't have to make 1,000 lighting pages.)

Don't overload your user by showing ALL THE THINGS.
The top half of the page should be dedicated to what a user needs to do NOW.

Are you asking the user for information?
Ask only for what's needed NOW.

Does the user need to complete several actions?
Which comes first? Show only that one. It's called progressive disclosure and your users will love you for it.

But the business says we need to show ALL THE THINGS.
If users still need to access all of the other 595 fields, chunk them up and put each chunk in its own tab under the main area we just talked about Not only with that reduce cognitive overload, it'll also make your pages load faster.

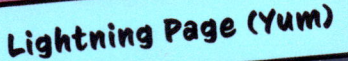

Lightning Page (Yum)

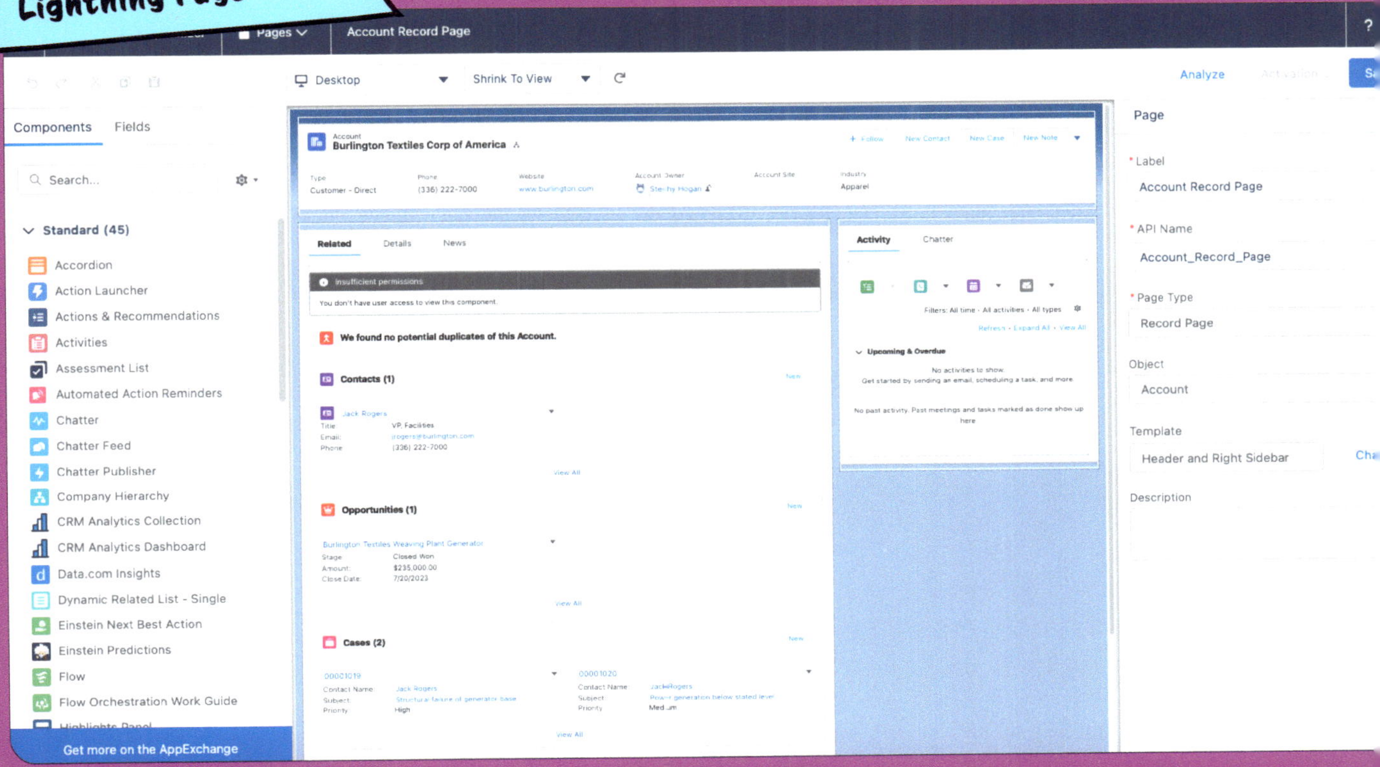

Desktop   Shrink To View   Analyze   Se

Components   **Fields**

Q Search...

∨ Standard (45)

- Accordion
- Action Launcher
- Actions & Recommendations
- Activities
- Assessment List
- Automated Action Reminders
- Chatter
- Chatter Feed
- Chatter Publisher
- Company Hierarchy
- CRM Analytics Collection
- CRM Analytics Dashboard
- Data.com Insights
- Dynamic Related List - Single
- Einstein Next Best Action
- Einstein Predictions
- Flow
- Flow Orchestration Work Guide
- Highlights Panel

Get more on the AppExchange

Account
Burlington Textiles Corp of America ∧

+ Follow   New Contact   New Case   New Note

Type: Customer - Direct   Phone: (336) 222-7000   Website: www.burlington.com   Account Owner: Sterhg Hogan   Account Site: Industry: Apparel

Related   Details   News

⚠ Insufficient permissions
You don't have user access to view this component.

⚠ We found no potential duplicates of this Account.

Contacts (1)   New

Jack Rogers
Title: VP, Facilities
Email: jrogers@burlington.com
Phone: (336) 222-7000

View All

Opportunities (1)   New

Burlington Textiles Weaving Plant Generator
Stage: Closed Won
Amount: $235,000.00
Close Date: 7/20/2023

View All

Cases (2)   New

00001019
Contact Name: Jack Rogers
Subject: Structural failure of generator base
Priority: High

00001020
Contact Name: JackRogers
Subject: Power generation below stated level
Priority: Medium

View All

Activity   Chatter

Filters: All time · All activities · All types
Refresh · Expand All · View All

∨ Upcoming & Overdue
No activities to show.
Get started by sending an email, scheduling a task, and more.

No past activity. Past meetings and tasks marked as done show up here.

Page

* Label
Account Record Page

* API Name
Account_Record_Page

* Page Type
Record Page

Object
Account

Template
Header and Right Sidebar

Description

Honestly, Salesforce gets a bad rap. Salesforce is immensely powerful and is improving constantly. As I tell all of my design friends...it's not the tool. It's the tool behind the tool. You don't blame the hammer if you build a chair with 2 legs.

Screen flows make no-code complex journeys possible.

What used to have to be custom-built components might not have to be anymore. So, even on the most basic of Salesforce projects, UX designers can make a HUGE impact without breaking the budget.

I'm not going into depth about how you can build a screen flow. But I DO want you to have a basic understanding of how they work and what they're capable of.

The biggest takeaway is that a screen flow lets you progressively disclose information and form fields to the users in an out-of-the-box way.

The user isn't overwhelmed by 876 fields at once.

No custom code is needed (unless you want a front-end developer to skin it).

The client saves time and money while increasing adoption and efficiency.

It's a win-win-win.

This will make sense in a second.

Let's say I want to take the user on a journey through their candy preferences. I might want to know things like:

- What kind of chocolate do you like?
- What do you pair your chocolate with?
- Do you even think of chocolate as candy?
- How do you feel about gummy bears?

If the user doesn't like chocolate, then I'm not going to ask them what they pair it with. I'll ask about candy.

If they don't like pineapple Haribo gummy bears, I might kick them out of the form altogether and deny them access to the club...and then post their banned status in a Slack channel.

Conditional forms are best practice.

You can take the user through this line of interrogation without someone writing code. You just need a declarative engineer. This is where I'm going to start using the term "Declarative UX."

The flow itself might look something like the diagram below. But the user will see... (BOOP! Time to turn the page).

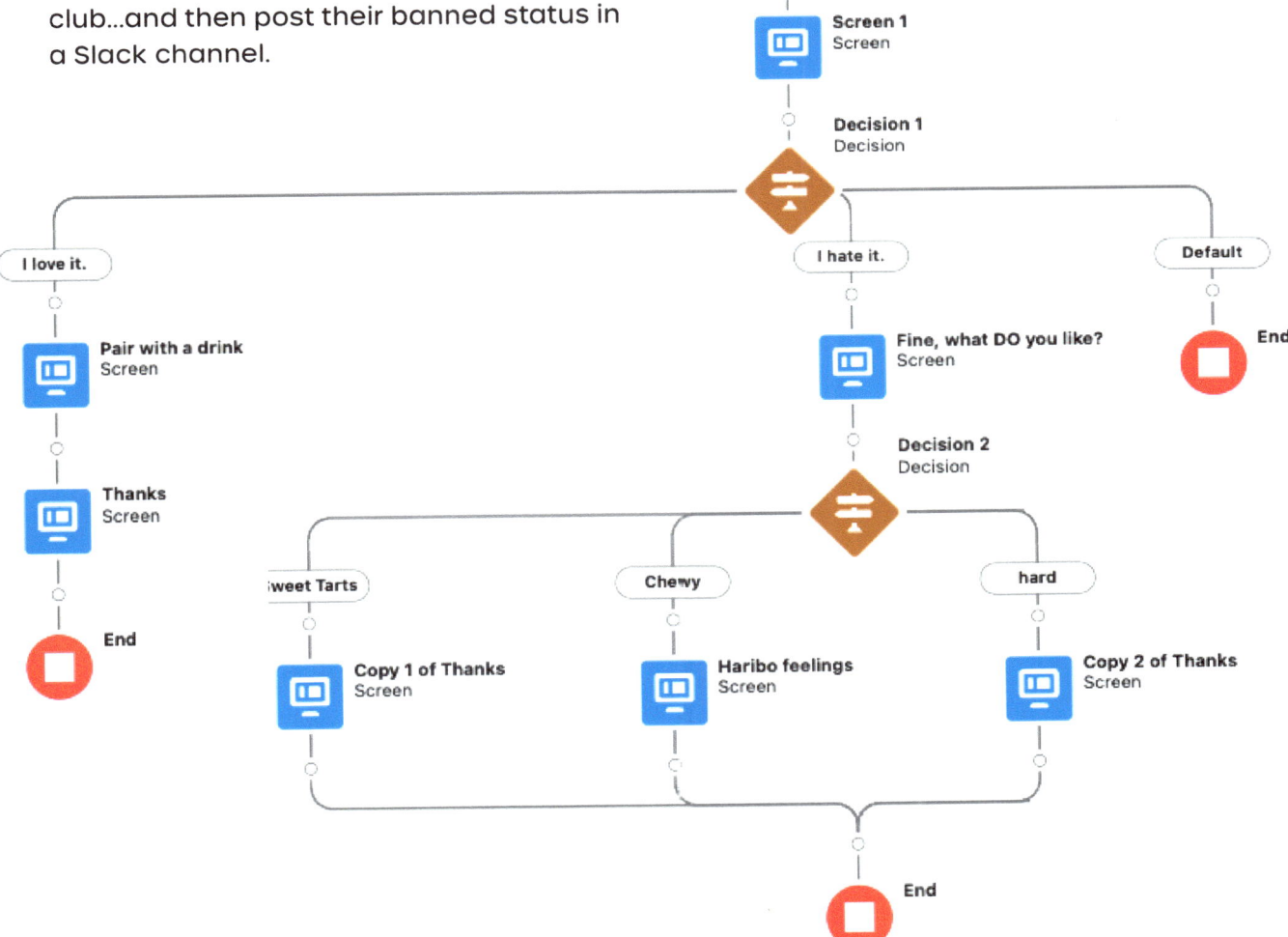

...a journey like this

First name

[   ]

Do you like chocolate?

(●) Yes   ( ) No

Do you pair it with scotch?

( ) Yes   (●) No

Thanks! Your Monopoly money is in the mail.

...instead of this

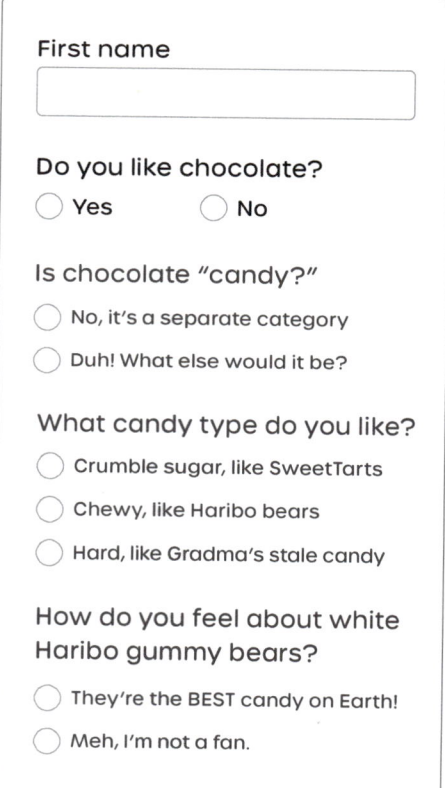

First name

[   ]

Do you like chocolate?

( ) Yes   ( ) No

Is chocolate "candy?"

( ) No, it's a separate category

( ) Duh! What else would it be?

What candy type do you like?

( ) Crumble sugar, like SweetTarts

( ) Chewy, like Haribo bears

( ) Hard, like Gradma's stale candy

How do you feel about white Haribo gummy bears?

( ) They're the BEST candy on Earth!

( ) Meh, I'm not a fan.

The screen flow might look like...

First name

[   ]

Do you like chocolate?

( ) Yes   ( ) No

if yes...

Do you pair it with scotch?

( ) Yes   (●) No

Thanks! Your Monopoly money is in the mail.

AUTHOR'S NOTE:

Ok, ok. This example doesn't look all that bad when it's in one form. But imagine we're building a college application form. Or a mortgage application process.

There is no conditionality in the single page form on the left. Yes, you CAN have conditionality in a one-pager. But it just adds to the user's cognitive load and decreases the pages performance.

Is chocolate "candy?"

- ● No, it's a separate category
- ○ Duh! What else would it be?

I completely agree! We can continue this survey.

Is chocolate "candy?"

- ○ No, it's a separate category
- ● Duh! What else would it be?

No. Chocolate is NOT candy. Let's continue.

What candy type do you like?

- ○ Crumble sugar, like SweetTarts
- ● Chewy, like Haribo bears
- ○ Hard, like Gradma's stale candy

How do you feel about white Haribo gummy bears?

- ○ They're the BEST candy on Earth!
- ○ Meh, I'm not a fan.

How do you feel about white Haribo gummy bears?

- ● They're the BEST candy on Earth!
- ○ Meh, I'm not a fan.

You're brilliant. I shall send you 10lbs of white bears.

How do you feel about white Haribo gummy bears?

- ○ They're the BEST candy on Earth!
- ● Meh, I'm not a fan.

You are wrong. Get out.

Peace out.

Dynamic forms are vital to a great experience.

Dynamic forms were first introduced in 2018 at TrailheaDX as "Lightning Layouts". Six years later, they're more powerful than ever – but still horribly underused.

A new friend of mine, Xavier Lisinski, is a data wiz. Using his superpower, he discovered that the average number of fields the user sees at the same time is SEVENTY EIGHT!

Imagine going to update a sales opportunity only to be presented with that many fields. Good luck finding the ones you need.

Seventy eight. SMH. (Not nearly the worst, my friend Michael Philbrick says he's seen some companies max out at 800.)

Hardly anyone is using dynamic forms to present only the fields a user needs at any given time or step in the process. It's a shame. Progressive disclosure is SUCH an important thing to practice in form design. Not only do we, as designers, like making things that aren't ugly and cluttered...users like not having their brains go to static when the see 78 fields in front of them.

If you're noodling around with editing Lightning pages and see this, ALWAYS HIT THE BUTTON. Everything should be a dynamic form. You're not going to break anything.

> **ⓘ Upgrade to Dynamic Actions**
> You can now configure dynamic actions for the highlights panel in the Lightning App Builder. The actions will be enabled for desktop.
>
> Upgrade Now

Just imagine if everything on a lighting page was dynamic...

A book publisher found a new writer that they want to work with and she enters in the writer's name, address, and genre.

Once the publisher chooses a genre (horror for example) a new field appears with a sub-genre picklist and text field for the author's favorite horror film.

Without dynamic forms, all of those fields would be present regardless of chosen genre— or that information would have to be entered on a separate screen.

27% of people say that form length is **THE** reason they abandon a form!

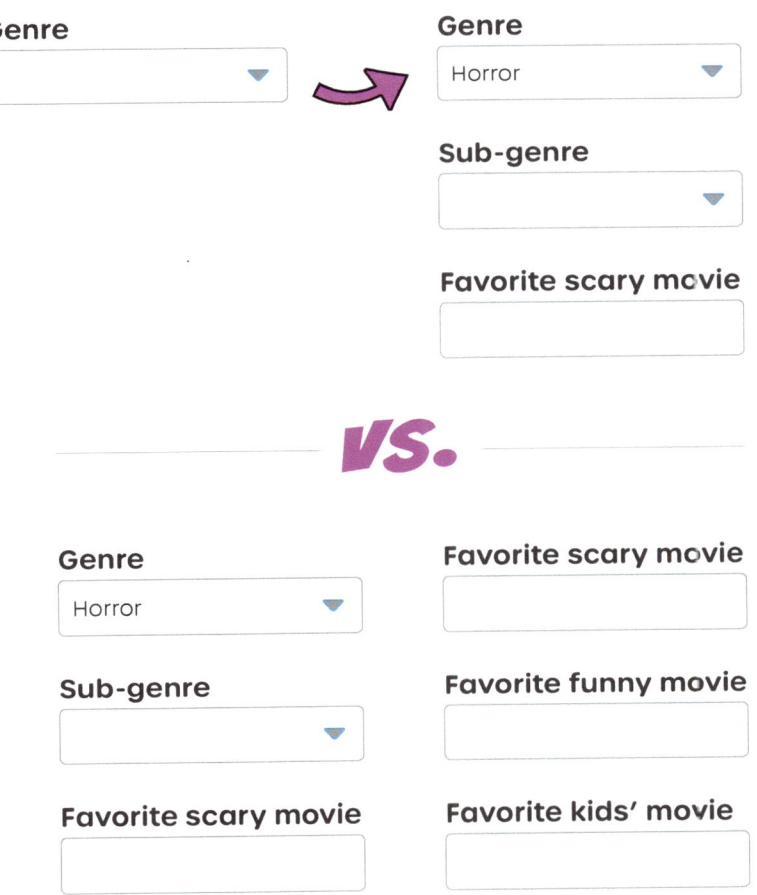

Genre

Genre

Horror

Sub-genre

Favorite scary movie

VS.

Genre

Horror

Favorite scary movie

Sub-genre

Favorite funny movie

Favorite scary movie

Favorite kids' movie

Need a real life example? Me too.

What can you do with all of those fancy things I just showed you? How can you help people not pull their hair out until they need to wear a hat so as not to blind their neighbors?

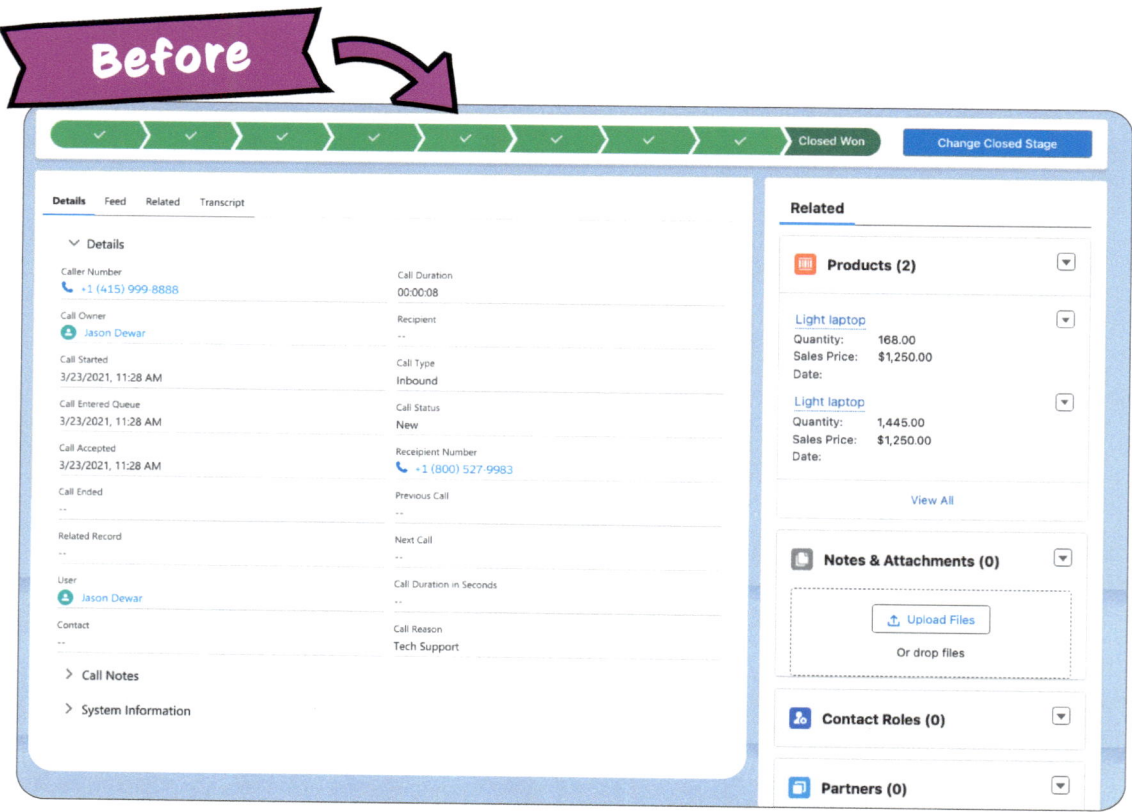

Here's a super simple example.

On the left is an out-of-the-box opportunity page. It doesn't have a lot going for it. Lots of stages! Lots of scrolling! (I mean, really, I spared you from seeing the fact that the sidebar is at least twice as long as the main panel.)

Down below, all three things (custom lightning page layouts, screen flows, and dynamic forms) have been applied. Each stage has its respective required fields directly below it. Anything else is moved to the tabs below it. There's user guidance! Imagine that. And any fields the user doens't use or need to see aren't shown anywhere at all.

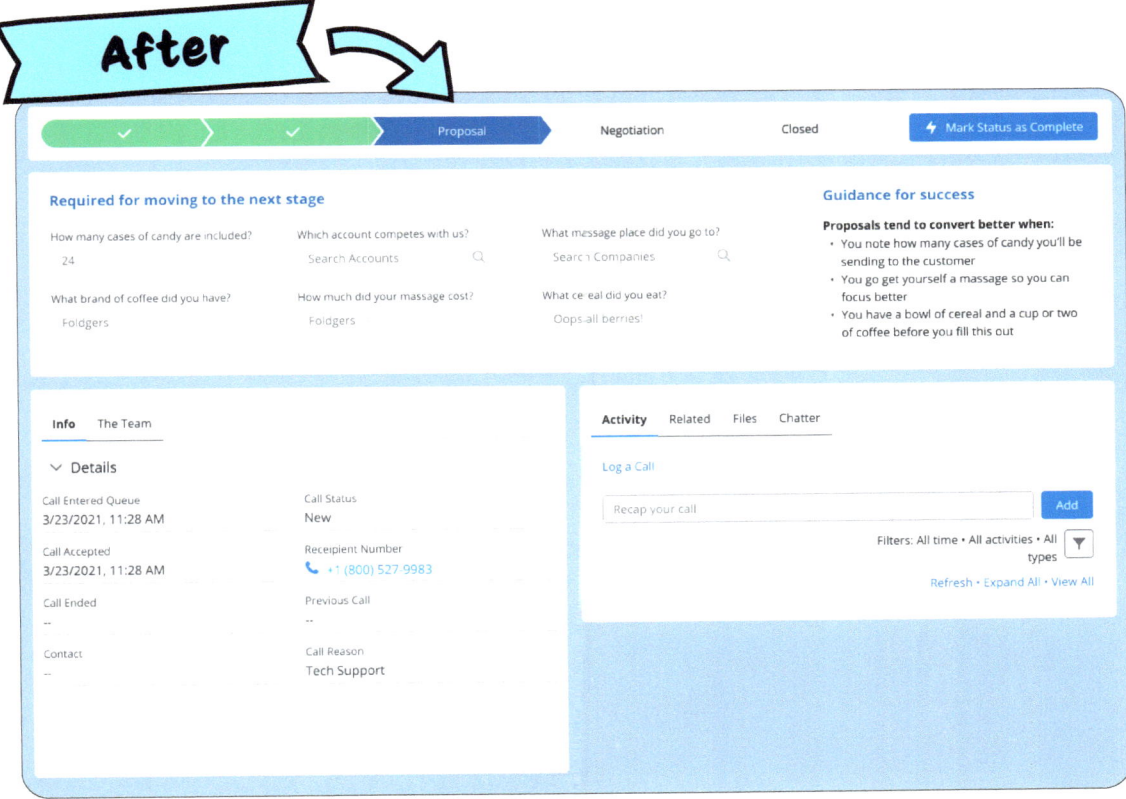

Can forms be better in Salesforce?

And when I ask that, it's a jab at the fact that Salesforce is basically nothing BUT forms. If we follow best UX practices for forms, it'll be one step closer to a usable system.

In our quest to create something that is super easy for the user to fill out, you'll come up against some roadblocks. Salesforce doesn't offer you all of the inputs you might need. And what fields ARE available to you also depends on if you're working straight out-of-the-box or creating custom components.

If you're creating custom, you'll almost always use the **Salesforce Lightning Design System** as the base, no matter how it's styled. So let's start there.

(Of course this design system is more than just that. And it's not bad. But the Figma file is terrible, at least at the time I'm writing this.)

lightningdesignsystem.com/tools/overview

Source: financesonline.com/form-abandonment-statistics

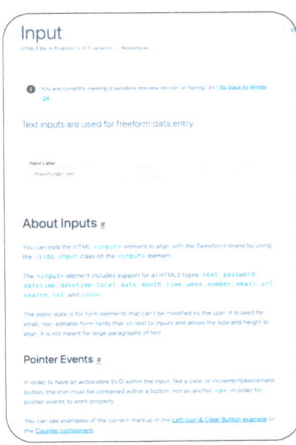

The documentation on the text input alone is lengthy.

Here is most of it broken up:

67%

percent of visitors abandon a form FOREVER if they encounter complications!

51

SLDS Form Elements

Some seem like dulicates and they kinda are.

Counter

Have your users count their clicks. They won't like it.

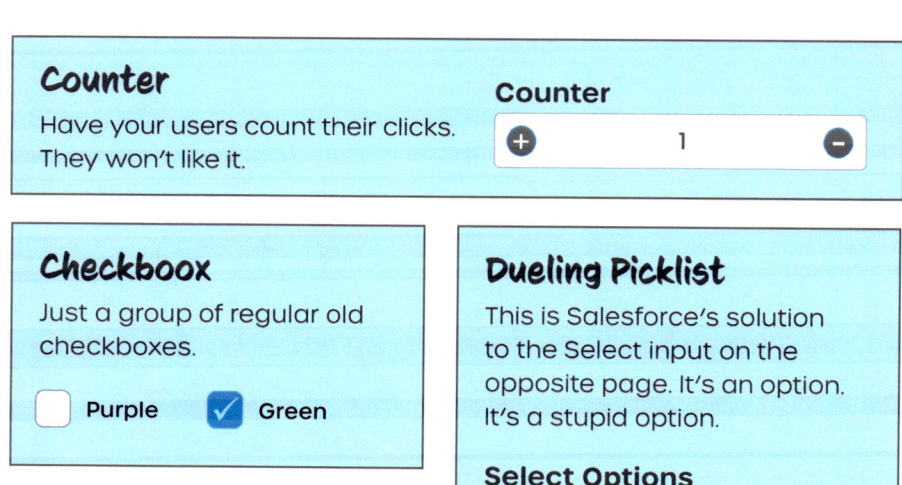

Counter

+ 1 −

Checkboox

Just a group of regular old checkboxes.

☐ Purple ☑ Green

Radio Buttons

Just a group of buttons

◉ Never gonna give you up
○ Never gonna let you down

Dueling Picklist

This is Salesforce's solution to the Select input on the opposite page. It's an option. It's a stupid option.

Select Options

Sequins	▶	Rhinestones
Glitter		
Feathers	◀	

Inputs (the text varietal)

Freeform text inputs... single line, text area, and rich text.

Input

Placeholder text...

Textarea

Placeholder text...

Rich Text

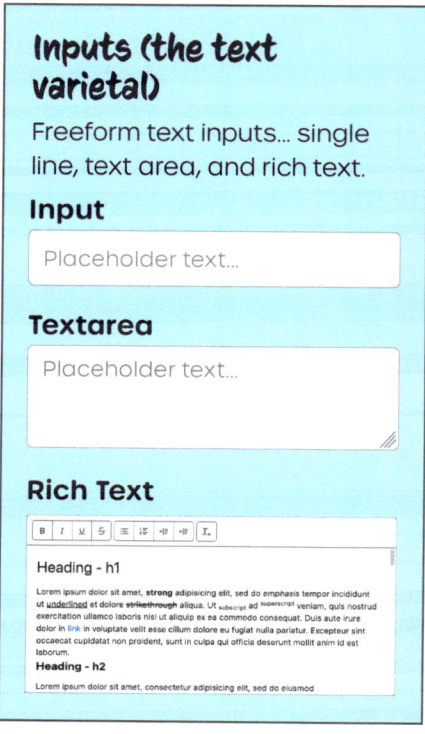

Heading - h1

Lorem ipsum dolor sit amet, **strong** adipisicing elit, sed do emphasis tempor incididunt ut underlined et dolore strikethrough aliqua. Ut subscript ad superscript veniam, quis nostrud exercitation ullamco laboris nisi ut aliquip ex ea commodo consequat. Duis aute irure dolor in link in voluptate velit esse cillum dolore eu fugiat nulla pariatur. Excepteur sint occaecat cupidatat non proident, sunt in culpa qui officia deserunt mollit anim id est laborum.

Heading - h2

Lorem ipsum dolor sit amet, consectetur adipisicing elit, sed do eiusmod

Slider

You might know this better as a Segmented Control, the multi-select version.

Slider Label
0 – 100

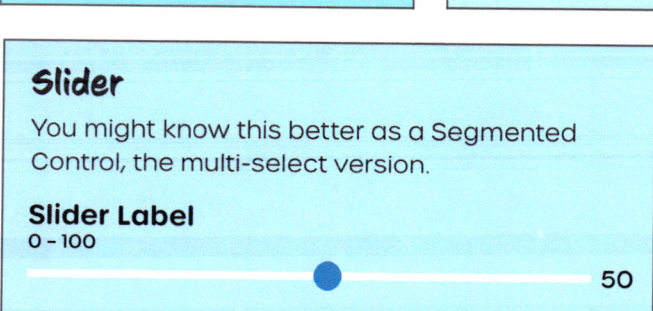

50

File Selector

Add pics of chocolate to your blog posts.

Upload Files or Drop Files

Checkbox Toggle

Turn lightswitch raves on and off.

Wifi

Enabled

Date/Time, Time, and Date Pickers

Combine the pickers or use separately as needed.

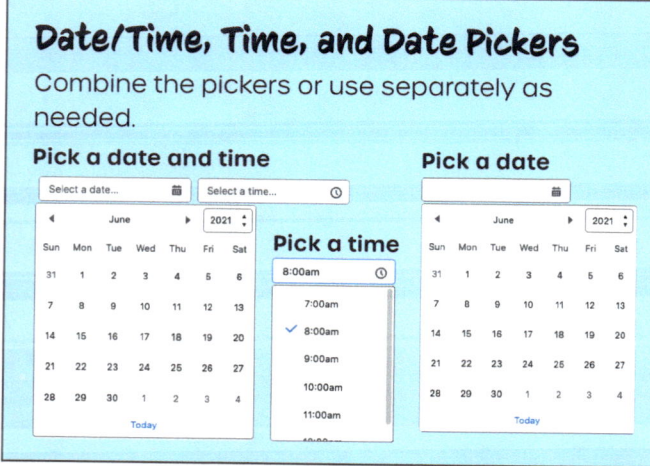

Visual Selector

Acts as a checkbox group, but it looks much fancier.

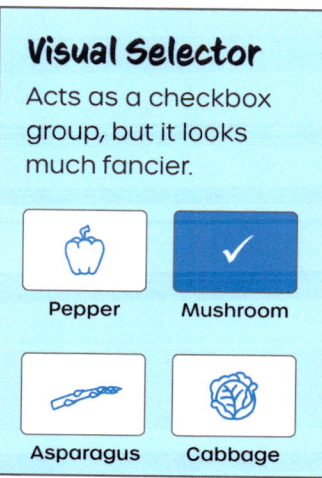

Color Picker

Just a group of buttons

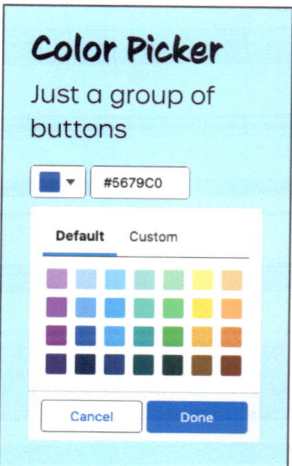

THIS mess...

When I'm in Figma and need a dropdown from the SLDS library, I always look for "dropdown" and I'm always disappointed. The SLDS website has all of these very similar looking items. They aren't all in Figma. And it doesn't matter.

Picklist

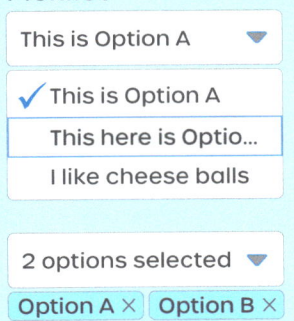

Salesforce says picklist, I say dropdown. You can allow users to **select only one** or **more than one** option. Picklists are **not their own component in Figma**. Use the combo box and use variants. If more than one is selected, this is what the field looks like.

Lookup

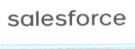

These are combo boxes that default to autocomplete and search against a database object (like Account or Contact). Lookups are **not their own component in Figma**. Use the combo box and use variants.

Combobox

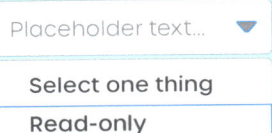

This is exactly like the select input. The only difference (that we care about) is that this one is SLDS style and the other is LEX UI style. This exists in the Figma file.

Select (default)

Select (open)

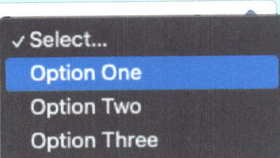

I honestly don't know why this is a thing if a picklist can do single or multiple selections. **Users can only select one option.** And the plural of a select field is a Dueling Picklist field. (ew)

...and THIS mess.

Buttons. Loads of buttons. These aren't even all of them with "button" in the name, just the one's you're most likely to use. I wish SLDS used words l ke "primary" and "secondary" instead of this...

Buttons

Brand Button
You would use this as the primary button.

Outline Brand Button
And this one would be the reversed primary.

Neutral Button
Maybe use this one as a secondary button?

Destructive Button
You could hijack this to make a different button.

Text Destructive Button
Same with this one.

Success Button
And this one. Or use as is. Whatever.

Button
It's a button, but looks like a link. With no underline. Why would I ever use that?

Button Group

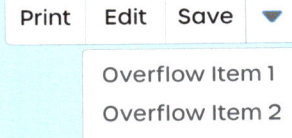

Just a group of buttons with no space between them. This is part of the general Salesforce interface.

Checkbox Button Group

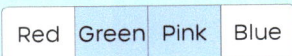

Also known as a multi-select segmented control.

Radio Button Group

Also known as a segmented control where the user can only choose one.

Lightning (LEX) Form Fields

You don't have a lot of options when it comes to out-of-the-box inputs in Salesforce Lightning.

You get fewer fields, which is rather disappointing. No sliders, no counters, no visual selectors (that's the one I'm REALLY sad about)

It's also worth mentioning that, yes, there are loads of buttons that happen in LEX. However, out-of-the box, you really only get to set an action color and Salesforce does the rest in ways you probably don't want it to, but oh well.

Oh, and fun fact: if you see anything on screen that looks like it came from the 1990s, it's a relic from Classic.

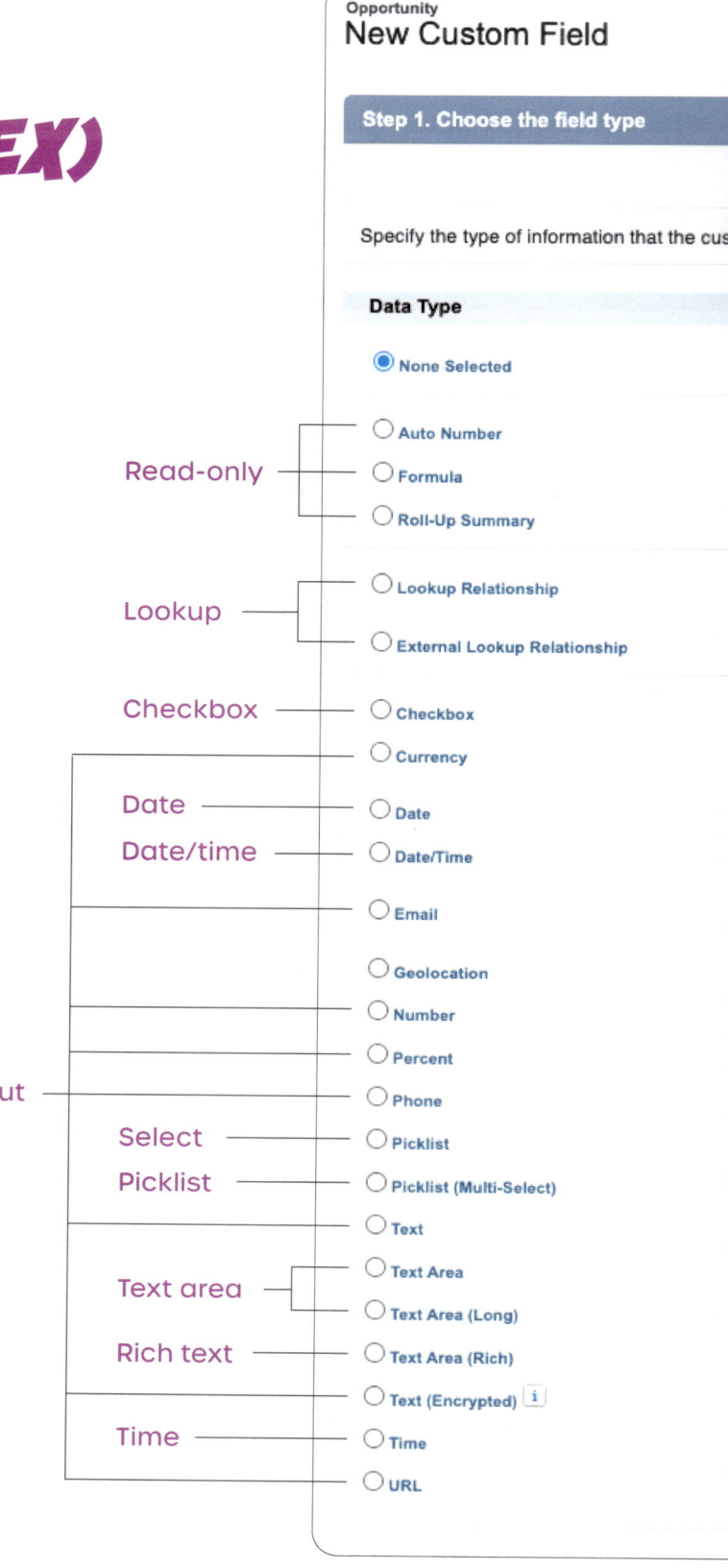

Lookup

This covers both Lookup Relationship and External Lookup Relationship

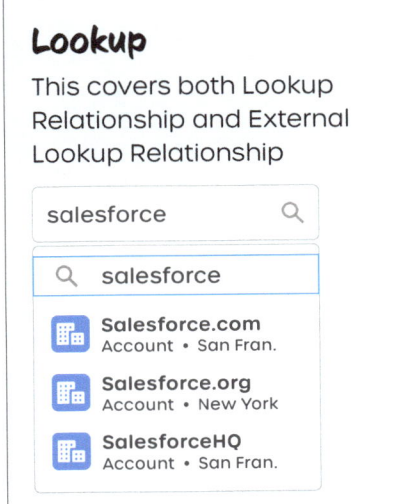

Checkbox

This covers checkboxes, both singular and plural.

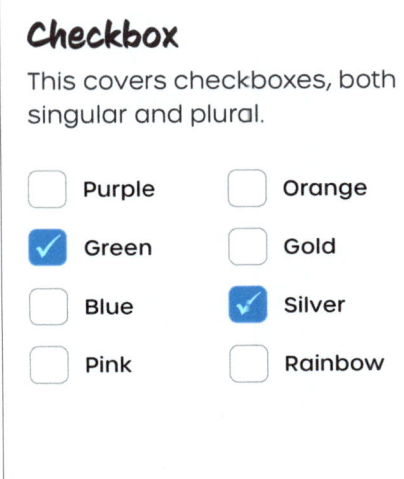

Picklist (single choice)

This is the sister field to SLDS's Select field.

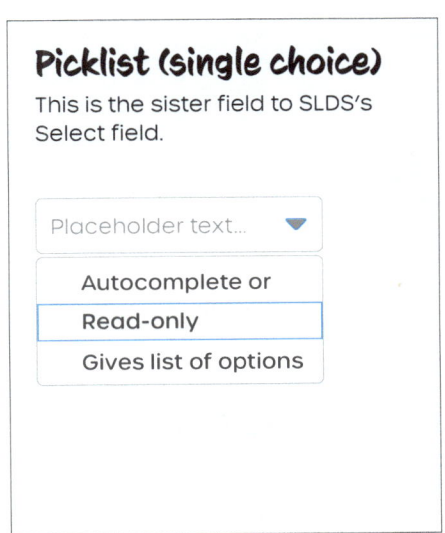

Date, Date/Time, and Time Pickers

Hey! Something that matches both SLDS and what we would normally expect as input fields.

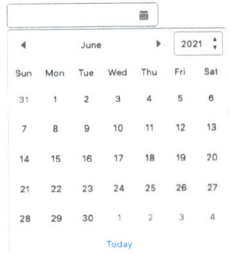

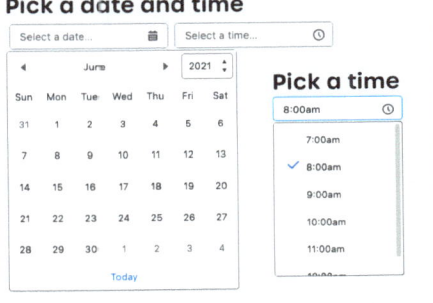

Radio Button

Never have just one.

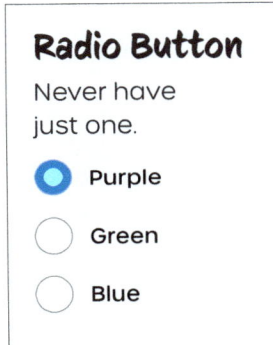

Picklist (multiselect)

In that input list on page 54, this is only called Picklist. In reality, it's a dueling picklist. Everyone says "dueling picklist." And they're dumb (the input, not the people).

Select Options

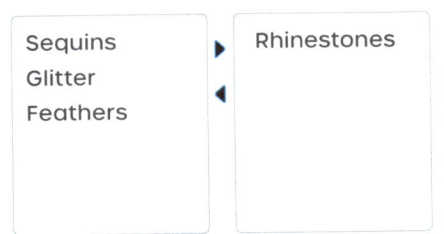

Inputs (the text varietal)

These are all just variations on a regular text input field. Lightning also adds in an excrypted text field which shouldn't affect how you design.

Input

Textarea

Rich Text

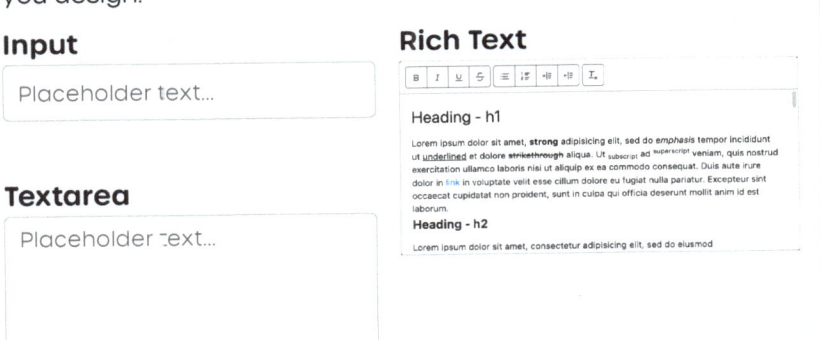

Screen Flow
Form Fields

And of couse, when you're building a screen flow, you have yet another different set of inputs to choose from. Because why the hell not.

Look on that page over there. --->

This is a bit nuts. Lemme 'splain.

You see a long-ass list of components and a tab for fields. The fields tab is empty by default and not worth explaining here since this isn't a screen flows manual.

These are the inputs available to you when you start making a screen flow. About half of them are really specialized for things within the org (all that order management, review, select stuffs). They're useful, but I don't count them as actual inputs in the context of this book so they're greyed out.

There are SLDS inputs that neither screen flows or LEX has:

- counter
- time
- visual selector
- color picker
- select (but who cares?)
- combobox (but again, who cares?)
- checkbox button group
- radio button group

Screen Flows

	LEX	SLDS
Components Fields		

Search components...

Input (51)

Component	LEX	SLDS
Address	☐	☐
Call Script	☐	☐
Cancel Appointment	☐	☐
Checkbox	☑	☑
Checkbox Group	☑	☑
Choice Lookup	☐	☐
Currency	☑	☑
Data Table	☐	☐
Date	☑	☑
Date & Time	☑	☑
Dependent Picklists	☐	☐
Display Image	☐	☐
Email	☑	☑
Enhanced Message	☐	☐
File Upload	☐	☑
Long Text Area	☑	☑
Lookup	☑	☑
Multi-Select Picklist	☑	☑
Name	☑	☑
Number	☑	☑
Order Management Enter Pa...	☐	☐
Order Management Enter S...	☐	☐
Order Management Order P...	☐	☐
Order Management Product...	☐	☐
Order Management Progres...	☐	☐
Order Management Select S...	☐	☐
Order Management Select S...	☐	☐

Screen Flows

Component	LEX	SLDS
Password	☑	☑
Phone	☑	☑
Picklist	☑	☑
Radio Buttons	☑	☑
Reassign Service Appointm...	☐	☐
Review Inbound or Outboun...	☐	☐
Review Service Appointment	☐	☐
Select Appointment Invitee	☐	☐
Select Appointment Type	☐	☐
Select Service Appointment...	☐	☐
Select Service Resource	☐	☐
Select Service Resource an...	☐	☐
Select Service Resources	☐	☐
Select Service Territory	☐	☐
Select Work Type Groups	☐	☐
Selected Promotion Segme...	☐	☐
Service Appointment Confir...	☐	☐
Service Resource Availability	☐	☐
Slack Channel Selector	☐	☐
Slack Workspace Selector	☐	☐
Slider	☐	☑
Text	☑	☑
Toggle	☐	☑
URL	☑	☑

Looking for field structure consistency?

Tee hee. Let's quickly go over how input anatomy is different between SLDS and Lightning. Because, as usual, it gets a little weird.

In SLDS, you basically get the functionality/parts that a designer would expect to get:
- Label
- Required notation
- Tooltip
- Descriptive help text
- Placeholder text
- Customizable error text
- Prefixes and suffixes, whether text or icon
- Clear buttons
- Character counts
- Loading spinner
- Steppers or counters
- Custom input formats (email, URL, phone, etc.)
- Custom placement of error and help text

In Lightning, you get:
- Label
- Required notation
- Tooltip, but it's called help text
- Customizable error text
- Custom input formats

SLDS

So many options!

Field-level help

Label

Fixed text

Left icon

Placeholder text

Inline help

Error text

Clear button
or right icon

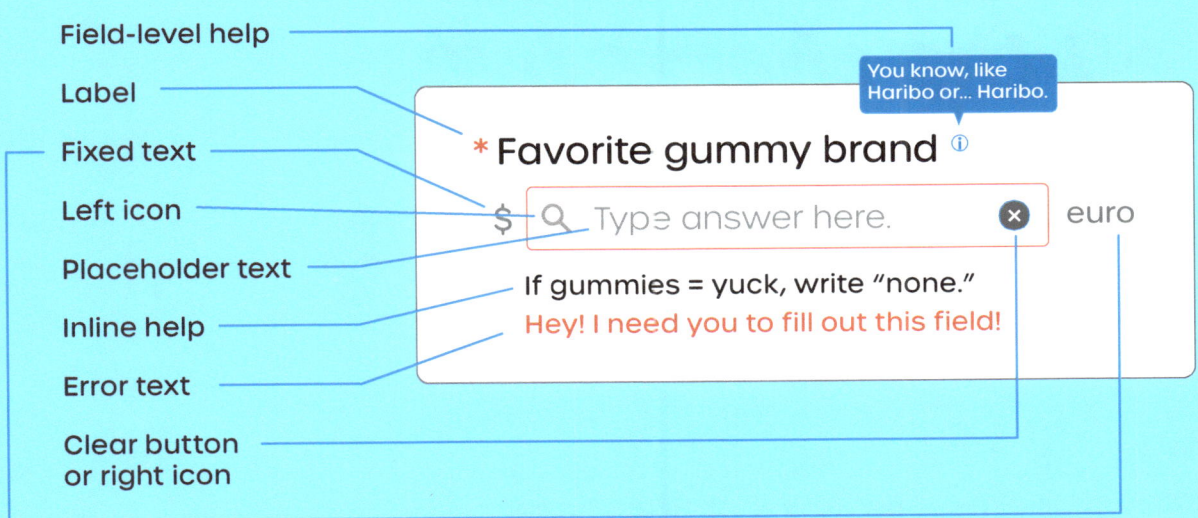

You know, like
Haribo or... Haribo.

* Favorite gummy brand ⓘ

$ Q Type answer here. ✕ euro

If gummies = yuck, write "none."

Hey! I need you to fill out this field!

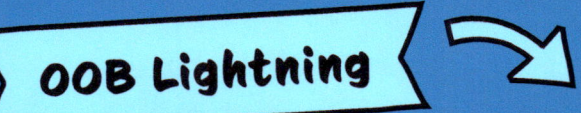

OOB Lightning

Yep, that's it.

Help text

Label

Validation text

You might not like
gummies. Write "none."

* Favorite gummy brand ⓘ

Hey! I need you to fill out this field!

Dashboards: Where all the charts live.

When Salesforce users aren't putting data into Salesforce, they're looking at the data already in it. I really love some beautiful data visualization. But I'm not going to get it from Salesforce.

Don't let the dashboard below fool you. Yes, it's actually pretty slick. When I first saw it, I thought, "Oh! Finally! Salesforce did something about the dated charting capabilities! This is gorgeous!

However, what you see on page 61 is only on sales-related homepages. It's called "Advanced Seller Home" and you can ether keep it on as-is or turn it off. If you turn it off, you have to make a new lightning page layout with a dashboard on it.

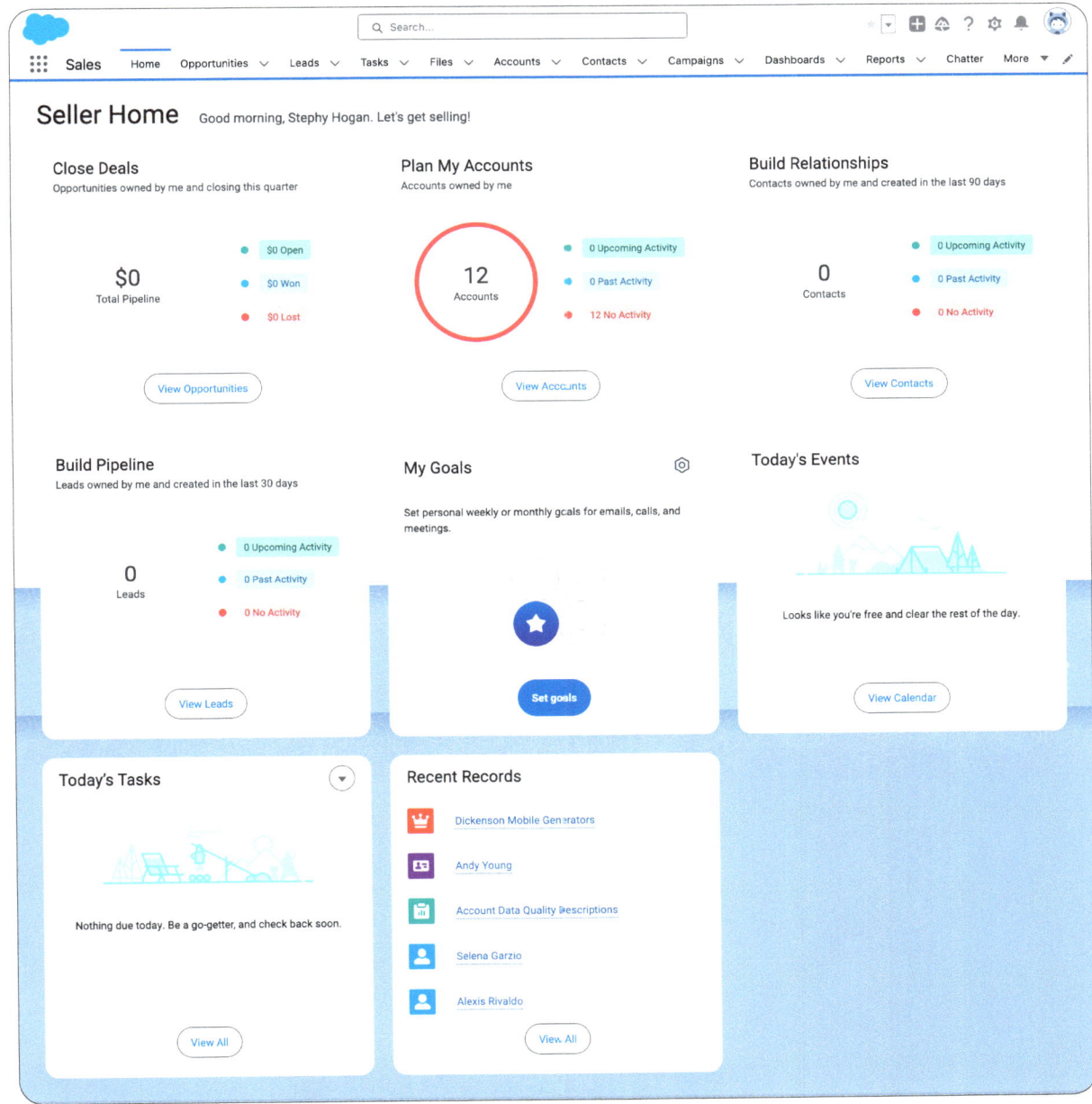

This is what it really looks like

Yep, down there. Check that out—the standard Salesforce charts. SIGH. And there's really not much you can do to make them prettier in the default dashboard or chart components. If you have some brilliant developers, they can make custom components and use your favorite charting library: D3.js, chart.js, etc. etc.

For now, all you can do is change the color theme the dashboard uses. At least there are light and dark themes. I guess.

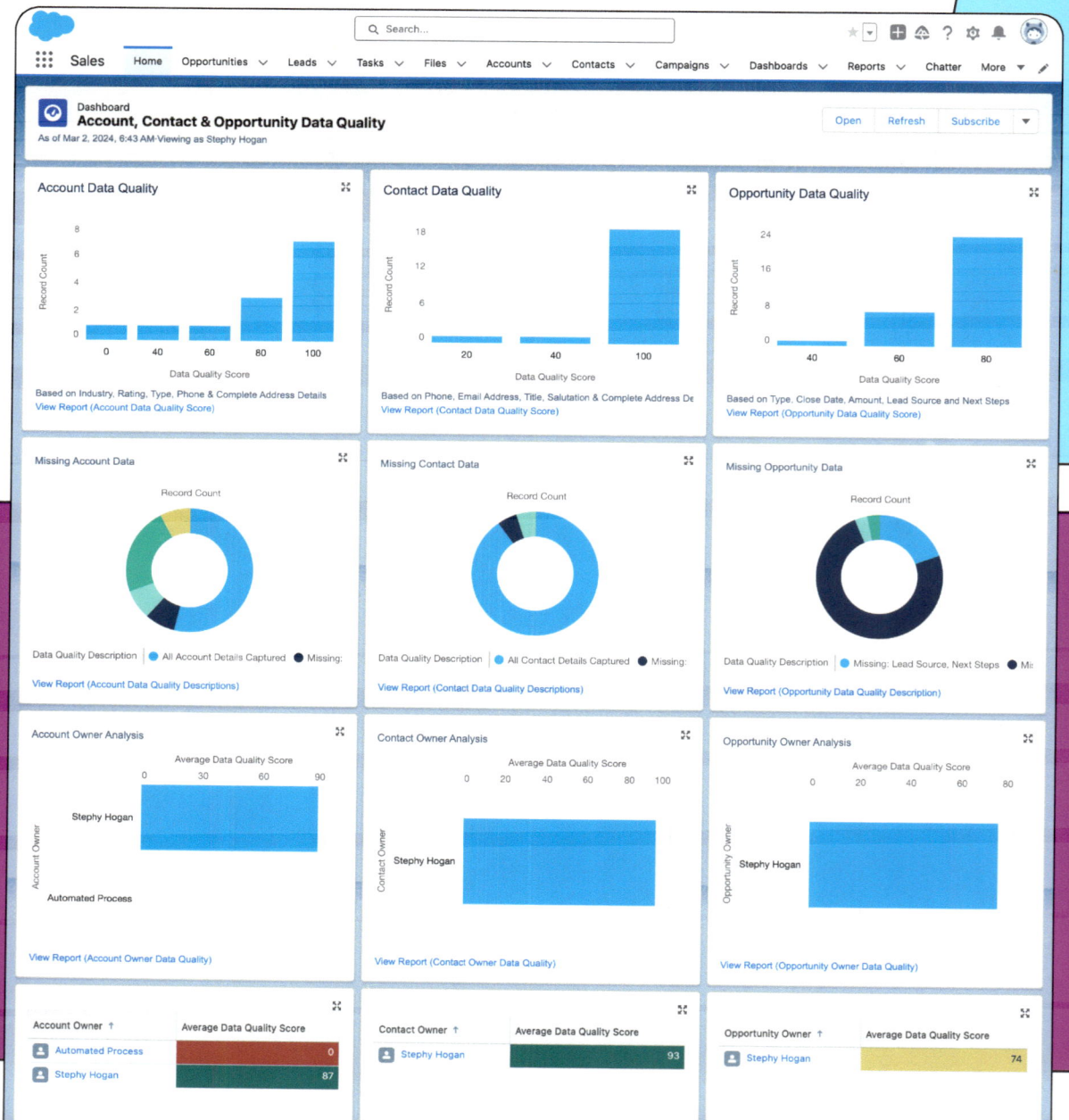

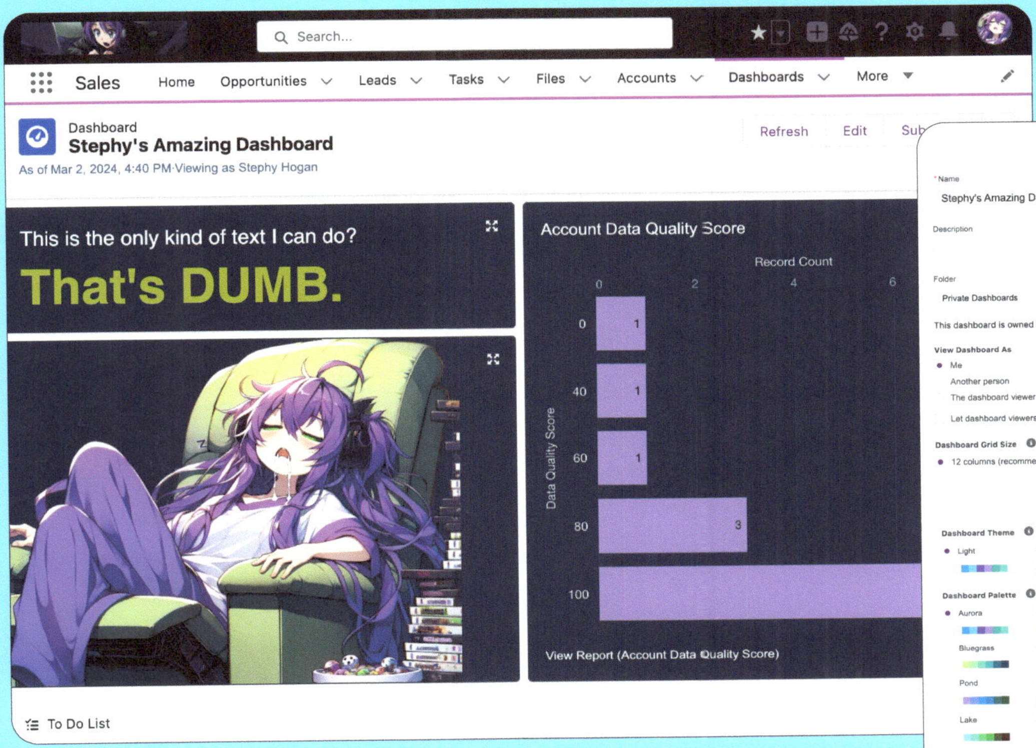

Dashboard screenshot

Sales Home Opportunities Leads Tasks Files Accounts Dashboards More

Search...

Dashboard
Stephy's Amazing Dashboard
As of Mar 2, 2024, 4:40 PM·Viewing as Stephy Hogan

Refresh Edit Sub

This is the only kind of text I can do?
That's DUMB.

Account Data Quality Score

Record Count

Data Quality Score: 0 → 1, 40 → 1, 60 → 1, 80 → 3, 100

View Report (Account Data Quality Score)

To Do List

You can add text blocks, images, and charts to the dashboard.

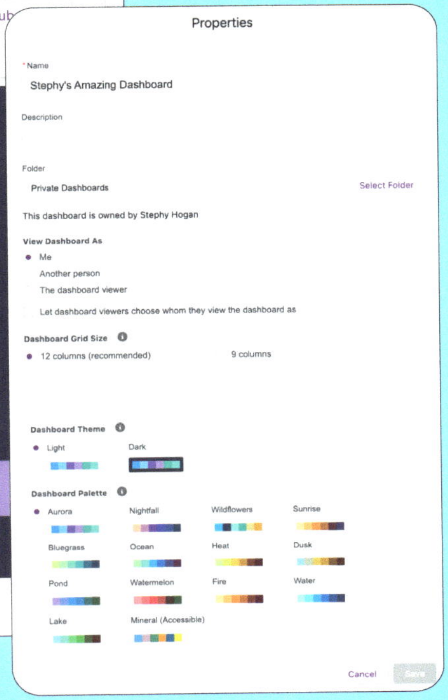

Properties

* Name
Stephy's Amazing Dashboard

Description

Folder
Private Dashboards   Select Folder

This dashboard is owned by Stephy Hogan

View Dashboard As
● Me
  Another person
  The dashboard viewer
  Let dashboard viewers choose whom they view the dashboard as

Dashboard Grid Size ⓘ
● 12 columns (recommended)   9 columns

Dashboard Theme ⓘ
● Light   Dark

Dashboard Palette
● Aurora   Nightfall   Wildflowers   Sunrise
  Bluegrass   Ocean   Heat   Dusk
  Pond   Watermelon   Fire   Water
  Lake   Mineral (Accessible)

Cancel   Save

Dashboard settings pane.

You get 10 chart types and a table.

The type of charts you can choose will depend cn the data being displayed. Bottom-left is a metrics chart and is ugly. And please, convince your stakeholders to stay away from pie charts.

Designing for a new org or an ancient one?

More often than not, you're going to end up making an old org or experience site that was set up in 2009 easier to use. Starting from a clean slate is a rare and much wished for situation. Why? Well...

Old orgs have a few problems:

- Custom components were built to perform functions that Salesforce does out-of-the-box these days.
- Lightning pages have hundreds more fields than they should because over the years, admins kept adding fields as the need arose.
- Dynamic forms are almost assuredly not being used.
- The data is a hot mess.

There are more, but I don't want to sit here get you down about what you've inherited. All of these issues have the potential to impact what MVP can be and what can actually be built, overall, taking budget, resources, and time into account.

So be prepared.

Expect these kinds of challenges.

Salesforce and accessibility

Salesforce has a lot of accessibility issues, both design and engineering related. They're improving, albeit slowly.

Lately, a load of contrast issues have been fixed, so yay! Now go check out this link and search for accessibility:

issues.salesforce.com

One of the things that bothers me the most is that if a company wants their Experience Cloud site to be perfectly accessible, you can't do it easily, if at all. And it's mostly because the shadow DOM (you'll learn about that on page 86) prevents engineers from customizing a lot of components to be accessible.

It makes me cranky.

There's a LOT to accessibility. There are ton of little nit-picky details you have to adhere to. The WCAG Guidelines are very comprehensive but BORING as shit to read. I'll let you choose do to that. Right now, I'll give you 5 types of users to consider. Please note the resulting acronym of the 5 together. Because: priorities.

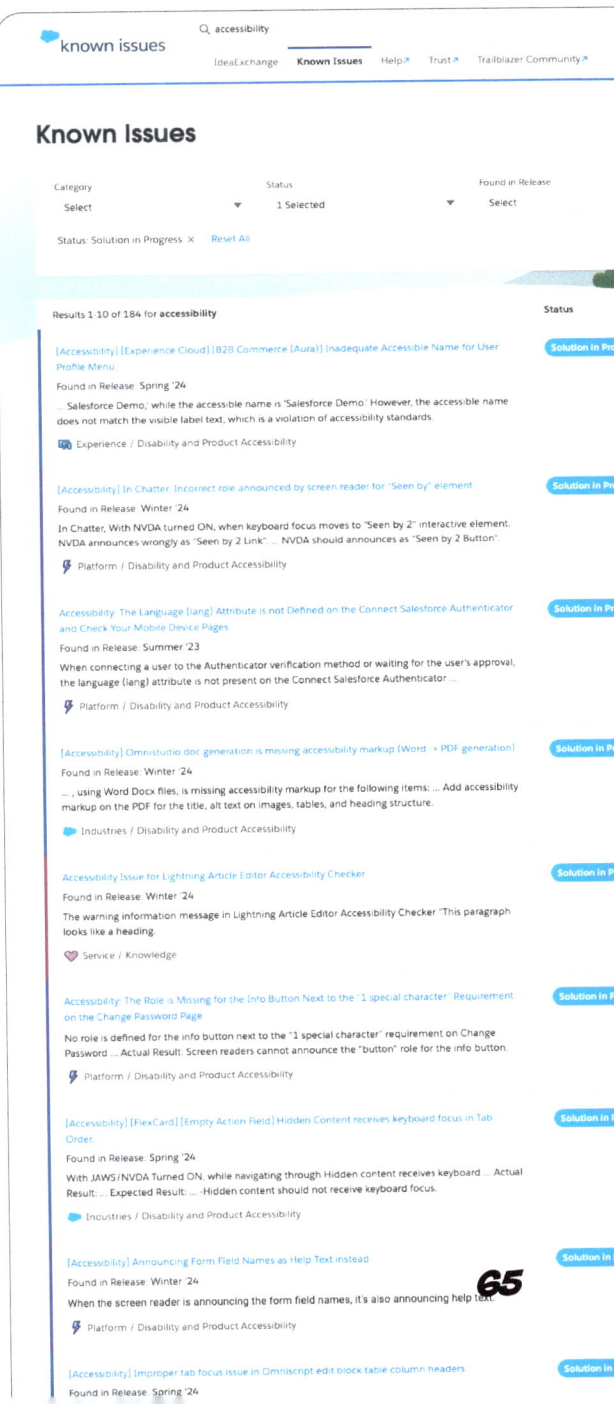

Sight-related accessibility

Can users see it? I'm not just talking color-blindness or total loss of sight...

Color contrast is where you're going to go first. It's the most visually impactful thing to fix and usually the most controversial (depending on what the brand colors are). In that regard, you'll probably get the most pushback from stakeholders when you start fiddling with colors. There are a few other things you need to remember about color usage both in and out of Salesforce. But who ARE you designing for to begin with?

The Who (not the band, but Tommy is totally my favorite musical):

- Sight Loss
- Color Blindness
- Cataracts
- Aging Eyes
- Light Sensitivity
- Screen Glare

Did you know that some people only see in grayscale?

The What:

COLOR CONTRAST

Yes, there's math, but you don't have to do any.

Color contrast is derived from comparing the luminance of each color in the pair. Luminance is the amount of visible light that is reflected off of (paper) or emitted from something (screens). Basic example? Black reflects little to no light. White reflects most or all light. If you're really curious, here's the math behind it.:

$$(L1 + 0.05) / (L2 + 0.05)$$

Isn't she pretty? So, if each of two colors reflects back about the same amount of light (regardless of color), it'll be really difficult to read because human eyes can't separate similar amounts of light very well. Physics is phun (a pun for all my science-minded friends out there).

There are a bunch of different tools you can use to check for contrast.

- Figma users: use the Stark plugin.
- Stark Chrome plugin
- AXE Chrome plugin
- WAVE Chrome plugin

If you can't install plugins, go to EightShapes Contrast Grid and enter hex values (be sure to include black and white)

Ratios to remember

Minimum contrast between text and the color it sits on. This is probably THE most important ratio to know. If you forget anything else, please remember THIS one. Yellow, orange, and green are THE trickiest colors to use successfully in any UI. If you have to use one of those colors, pay extra attention to how they're being used.

Minimum contrast between text larger than 18 points (24 pixels) and the color it sits on. Personally, I always aim for 4.5:1 no matter what size the text is. Why? Let's call it a safety net for those parts of a site where users enter content. The more users who think 4.5:1 is The Rule, then the chances of 10pt font paragraphs occurring are low.

Minimum contrast between interactive elements and the color they sit on. Things like buttons and inputs, for example, need to contrast enough from the background for users to know that they're there and that they're interactive.

Another fun fact is that there IS such a thing as too much contrast.

Wild, I know.

People who have light sensitivity issues have trouble with pure black on pure white and vice versa. People with autism can have issues with really saturated colors. Text will vibrate to those with Irlen Syndrome or other light sensitivity issues. Moderation is key.

Salesforce has somewhat recently fixed a TON of contrast issues in the interface. They improved colors for charts! They allow you to choose an accessible version of your brand color in themes! Pretty nifty.

However, there are still a lot of areas that need improvement. Charts are better, yes. However, no one has yet taken into consideration the contrast between adjacent chart colors. For example, no one can see the difference between the light blue and the dark teal donut slices in that chart over there. (The slice between the light blue and mint green slices.)

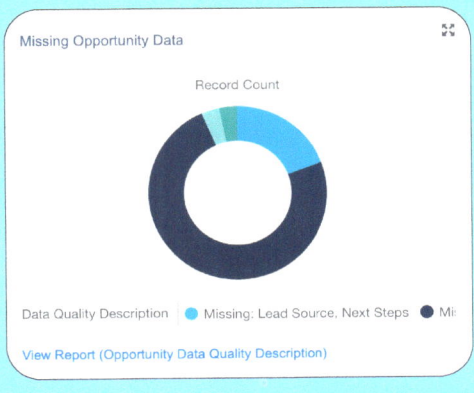

I hear a few of you saying, "But there's a legend!"

Friends. Lean in closely.

Just because there's a legend doesn't mean anyone can tell which label goes where when the same color issue exists in the legend. Also... this chart shouldn't be a donut chart and all of the bars should be directly labeled. Problem is that Salesforce doesn't allow you that much control. Here's to hoping there will be an update that fixes this in the future. This chart is also an excellent segue into the next point.

COLOR CAN'T BE THE ONLY INDICATOR OF MEANING

The way that chart is currently set up, color IS the only indicator of meaning. And this leads me into my second biggest pet peeve with the Salesforce interface. Links. Default Salesforce style is no underline–just a color change. No bueno, Salesforce. Color cannot be the only indicator of meaning. If you have the power and ability to do so, reader, whenever possible, get that underline in there.

FONT SIZE CONSIDERATIONS

How old are your eyes? What's the resolution of your screen? And can someone please tell me why Salesforce chose 13pt to be the default paragraph text size? Best practice is no smaller than 16pt for that kind of text–even input label and placeholder text. I don't think that day will ever come, so please, for me: don't go below the standard Salesforce size of 13pt for labels.

Fun fact... points and pixels aren't a 1:1 conversion.
1 point = 1.33 pixels.

Cognition-related accessibility

Can they UNDERSTAND it? Cognitive overload is often the most overlooked aspect of design and it's probably the one that affects the most people. Basically...stop putting *everything* on the page! You'll make people cry.

The Who:

- Humans with ADHD and other neurodivergencies
- People facing language barriers
- Users with dyslexia
- The sleep-deprived (Let's face it, that could be any of us on any given day.)
- Humans distracted by other humans (Ever tried to concentrate on a cluttered screen while someone's loudly discussing their weekend plans? Not fun.)

The What:

Now that you know who you're building for, let's talk about how you can make their lives easier (and yours, by reducing the number of helpdesk calls!).

SIMPLIFY YOUR LAYOUTS

A cluttered layout is the cognitive equivalent of a hoarder's basement. It's hard to find what you need in a mess. Keep your Salesforce interfaces as clean and organized as possible. Think Marie Kondo meets software design—if it doesn't spark joy (or at least clarity), toss it out!

When you chunk out what a person is required to fill out *and* put it at the top, even people having the foggiest of brain days will have an easier time getting the job done.

An example is the default opportunity page layout. This is where making use of personalized apps, better Lightning page layouts, dynamic forms and screen flows can really make a difference.

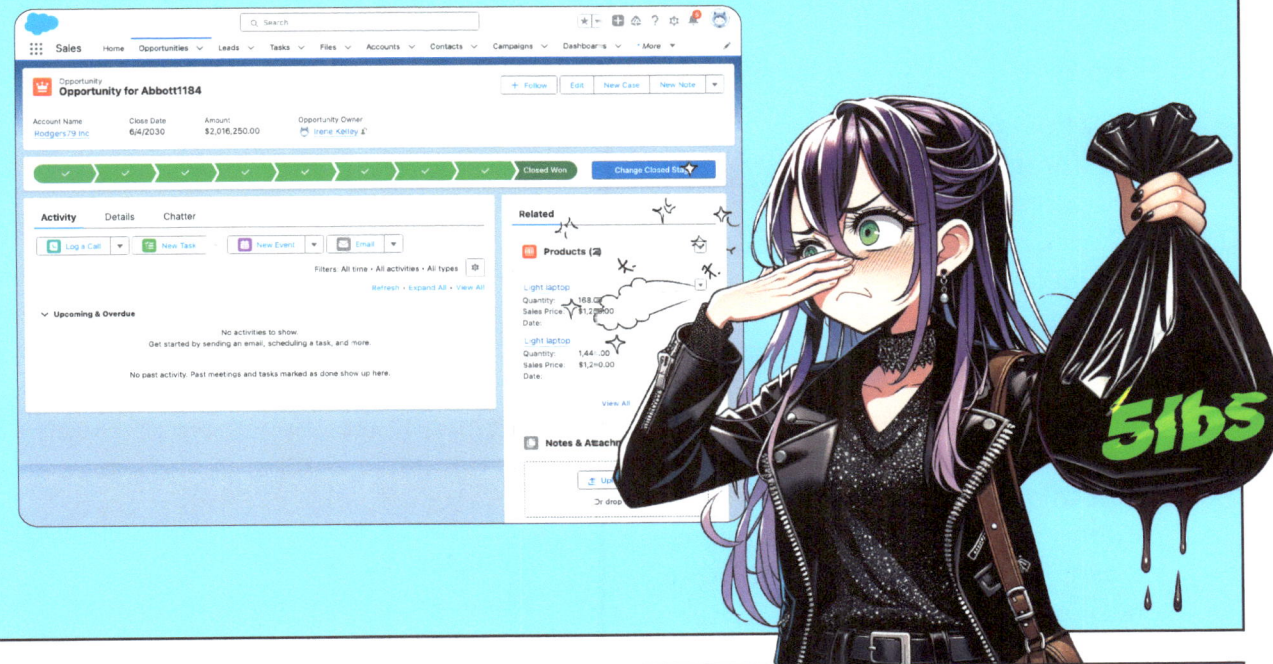

But by talking to your users to find out what they really need and then take a small amount of time to make a new layout, it could look more like this.

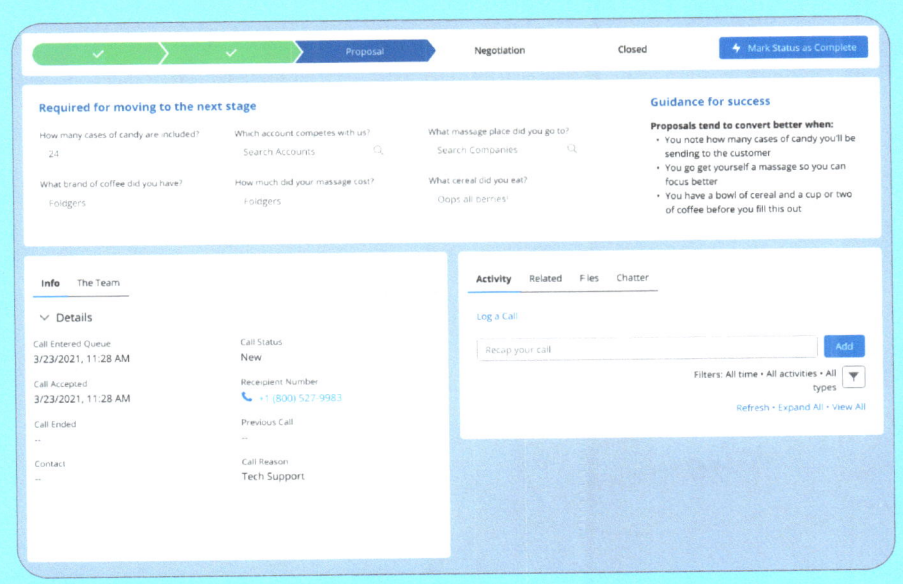

USE PLAIN ENGLISH

Use conversational language that's easy to understand. If you find yourself using phrases like "synergistic management solutions," it's time to drink less of that corporate Kool Aid.

If you haven't noticed already, Salesforce seems to have a consistency problem when it comes to the vocabulary they use throughout the whole ecosystem. Let's take care not to exacerbate the issue in field labels, object names, and app names.

DEMYSTIFY THE NEXT STEPS

Don't leave users hanging wondering what to do next. Always provide clear, actionable next steps. This could be as simple as a "Submit" button that changes to "Submitted!" to confirm the action. If the process the user is in is a lengthy multi-step one, provide information in each step about any materials the user will need to gather before moving on to the next step. Or provide insight as to what happens after that "Submit" button is clicked.

TEXT ALIGNMENT AND DECORATION

Keep your text alignment consistent and avoid using multiple font styles and sizes that can make the content harder to follow. Stick to one or two at most—like a black tie dress code but for your text.

Unless it's a button, please avoid centered text at all costs.

AVOID INFORMATION OVERLOAD (TMI)

Sometimes, less is more. Don't bombard users with all the data at once. Use progressive disclosure techniques to show more information as needed, not all upfront. Screen flows are fantastic in this regard.

EXPLAIN YOUR DATA

Make sure any data presented is clearly explained. No one should have to play detective to understand what they're looking at. And I don't mean going into a jargon filled diatribe about that data...tell the users why they should care and then provide the information in an easily digestible way.

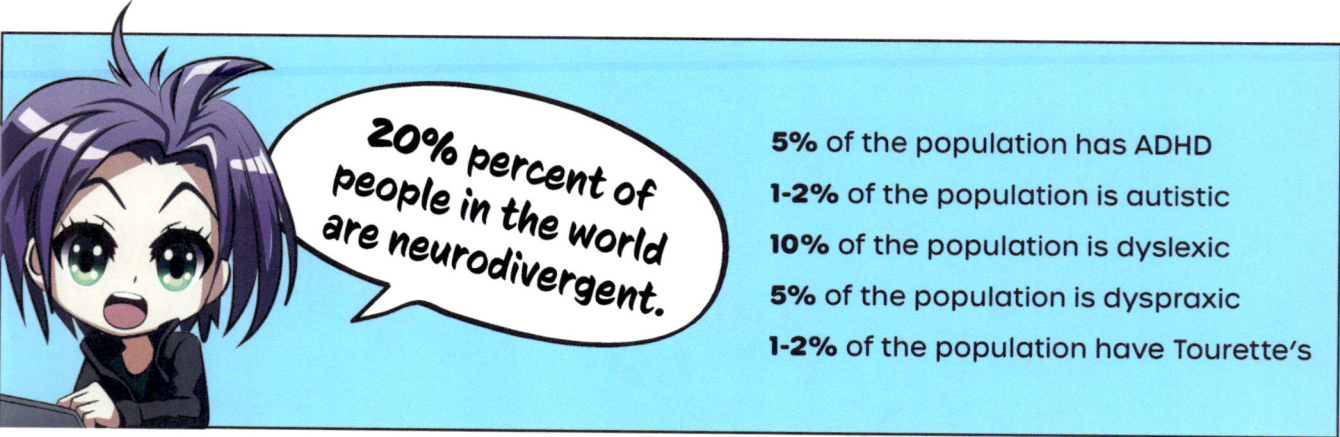

20% percent of people in the world are neurodivergent.

5% of the population has ADHD

1-2% of the population is autistic

10% of the population is dyslexic

5% of the population is dyspraxic

1-2% of the population have Tourette's

Source: Dr. Nancy Doyle, based on the work of Mary Colley

Hearing-related accessibility

Can they HEAR it?

The Who:

It's time to turn down the volume and focus on users who might not hear your audio cues at all. Consider these groups:

- Deaf humans
- Hard of hearing
- People in noisy environments
- Workers with noisy coworkers
- Victims of the dreaded leaf blowers

The What:

Understanding the challenges is one thing; designing solutions is where the real fun begins. Let's break down how to make Salesforce not only usable but enjoyable for everyone, regardless of their hearing ability.

ALT TEXT

Every image should tell a story, literally. Alt text helps users who use screen readers understand what they can't see. This is especially crucial for informative graphics.

CLEAR VISUAL SIGNALING

Lights on stage aren't just for show—they guide. Similarly, visual alerts (like wiggling icons) can signal new messages or updates.

ADJUSTABLE AUDIO CONTROLS

Give users the ability to turn down, turn up, or turn off.

DETAILED CAPTIONS & TRANSCRIPTS

Every video or audio file within Salesforce should have captions. Not only is this a lifeline for those who are hard of hearing, but it's also a blessing when someone's trying to catch up on training videos in the midst of the aforementioned leaf blowers.

Interaction-related accessibility

Can they INTERACT with it?

The Who:

When we talk about interaction, we're focusing on users who might face challenges simply navigating through what many of us take for granted. Here's who you should keep in mind:

- Individuals with Muscular Dystrophy
- People with Tremors
- Humans with Arthritis
- Older Adults
- Someone Missing a Limb
- People with a Broken Arm
- The Multitaskers

The What:

Let's break down the essential modifications and considerations to make Salesforce not just usable, but comfortably usable for everyone.

TAP (AND CLICK) TARGET SIZE

Enlarging tap and click targets within Salesforce's mobile interface or any web-based applications ensures that buttons and links are easy to press, even for those with limited dexterity or precision. This includes, but is not limited to:

- increasing the size of icons in the Salesforce App
- ensuring that menu items have sufficient spacing
- Ensuring text links aren't limited to single short words
- Ensuring the space between interactive elements is large enough to prevent fat-thumb syndrome.

The minimum size for these interactive elements AND the space between them is 24px by 24px for AA compliance.

KEYBOARD NAVIGATION

Ensure that every interactive element in Salesforce can be accessed with a keyboard. Make sure that keyboard navigation takes users through everything on screen in a logical order. Users who are listening to your website should be able to get the information they need just as quickly as those who aren't.

AVOID KEYBOARD TRAPS

It's crucial that users never find themselves stuck in a part of your application unable to move forward or back without using a mouse. This involves rigorous testing of new components and custom-built solutions in Salesforce to ensure that they can be exited or navigated through entirely with a keyboard.

ONE-HANDED USABILITY

Optimize Salesforce mobile apps for one-handed use, which not only helps someone holding a cup of coffee but also aids those with a temporary or permanent disability affecting one arm or hand. This could involve thoughtful placement of key actions within thumb reach.

Technology-related accessibility

Is the technlogy helping or hindering users?

The Who:

When we talk about technological accessibility, we're referring to all the versions of hardware, software, and connectivity scenarios that your users might encounter. Here's who you need to keep in mind:

- Users with Varied Hardware Quality
- Different Software Versions
- Varying Levels of Internet Connectivity
- Device Type

The What:

Ensuring that Salesforce performs optimally across all these different technological contexts involves a blend of thoughtful design and technical savvy. Let's explore how:

OPTIMIZE PAGE LOAD TIMES

Be cautious with how much you pack onto a single Lightning page. Hundreds of fields, complex layouts, and heavy components can significantly slow down the

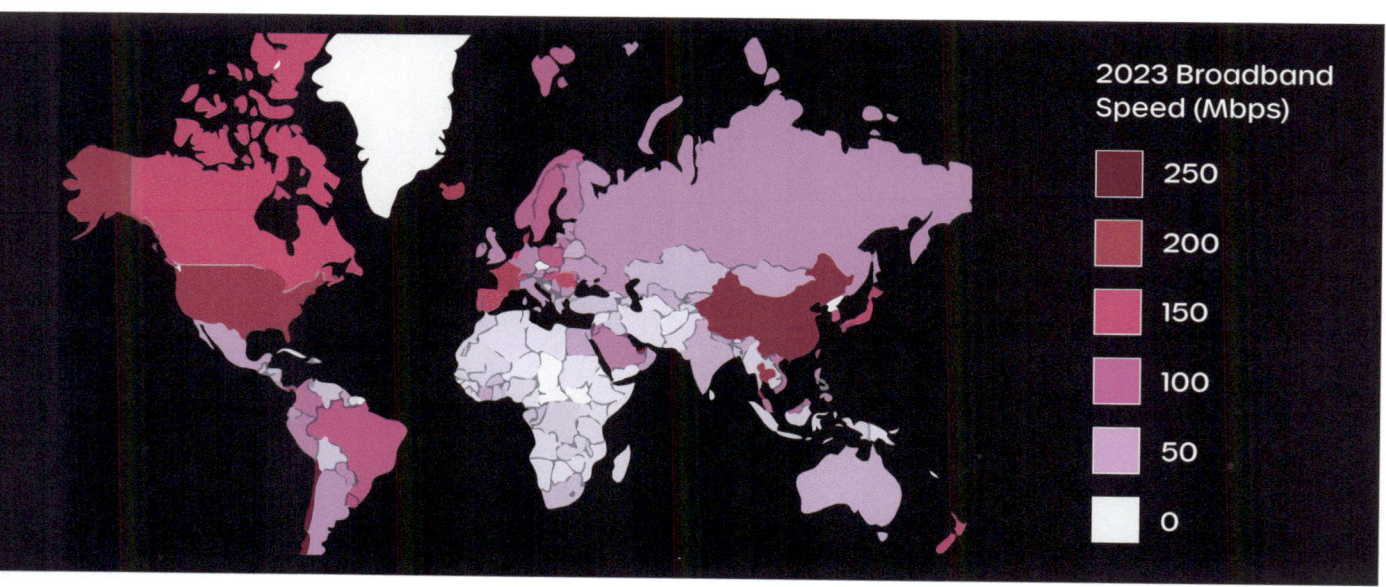

Source: worldpopulationreview.com

load time, particularly on older hardware or slower internet connections. Prioritize and streamline the essential elements to enhance performance.

MOBILE-FRIENDLY DESIGN

Ensure that your Salesforce Lightning pages and Experience sites are responsive, meaning they adjust smoothly to different screen sizes. When elements on a page reflow for smaller screens, the order and logic should remain clear and intuitive. This is crucial for users on mobile devices, ensuring they can navigate and interact with your pages as easily as desktop users.

EVALUATE CUSTOM COMPONENTS

Custom components, especially legacy ones, can be a double-edged sword. While they offer tailored functionality, they might not be optimized for performance. Regularly review and update custom code to prevent these components from bogging down page speeds. Consider newer, lighter-weight alternatives that achieve the same goals with better efficiency.

API MANAGEMENT

Integrating Salesforce with other systems via APIs adds functionality but can also introduce latency. If your setup requires pulling data from multiple sources, monitor and optimize these processes to ensure they don't hinder overall system performance. Sometimes, less is more— or at least, well-configured is more.

Get yourself a fat marker and write the acronym here. Don't be shy.

27%
percent of people in the world are living with some form of disability.

Source: cdc.gov

And now it's time.

Connect the gummy bears.
You'll never guess what it is.

Prepare yourself.

(my favorite)

Designing for Experience Cloud

Experience Cloud is my favorite place to be, especially when the client has a big budget and we have the time to make the most amazing website ever.

You can do ANYTHING in Experience Cloud if you have a team of great developers and access to the users who will be using what you build.

When you find yourself on one of these projects, I cannot stress enough how important it is to think in a crawl, walk, run manner. While you want the client to start with "crawl," YOU need to start with "run." Run is the most ideal amazing version of the project—the superhero version. It exceeds the client's wildest dreams. It's gorgeous. The users LOVE it. It surprises and delights. And you won't know what the "crawl" version will look like unless you reverse engineer from the "run" version.

I have heard a lot of people say that if we're building an MVP, then only design the MVP. **That would be a grave mistake.** You need to design in a way that takes into account all of the custom, complex, weird journeys and functions. Your process *should* be something like this:

1. Establish basic functional needs
2. Start consulting with your devs
3. Design ideal state (the Run prototype)
4. Design an effective way for the client to get there in stages (Crawl, Walk, Run prototypes)

So let's get to know Experience Cloud and figure out how to bend its Salesforceyness to our will.

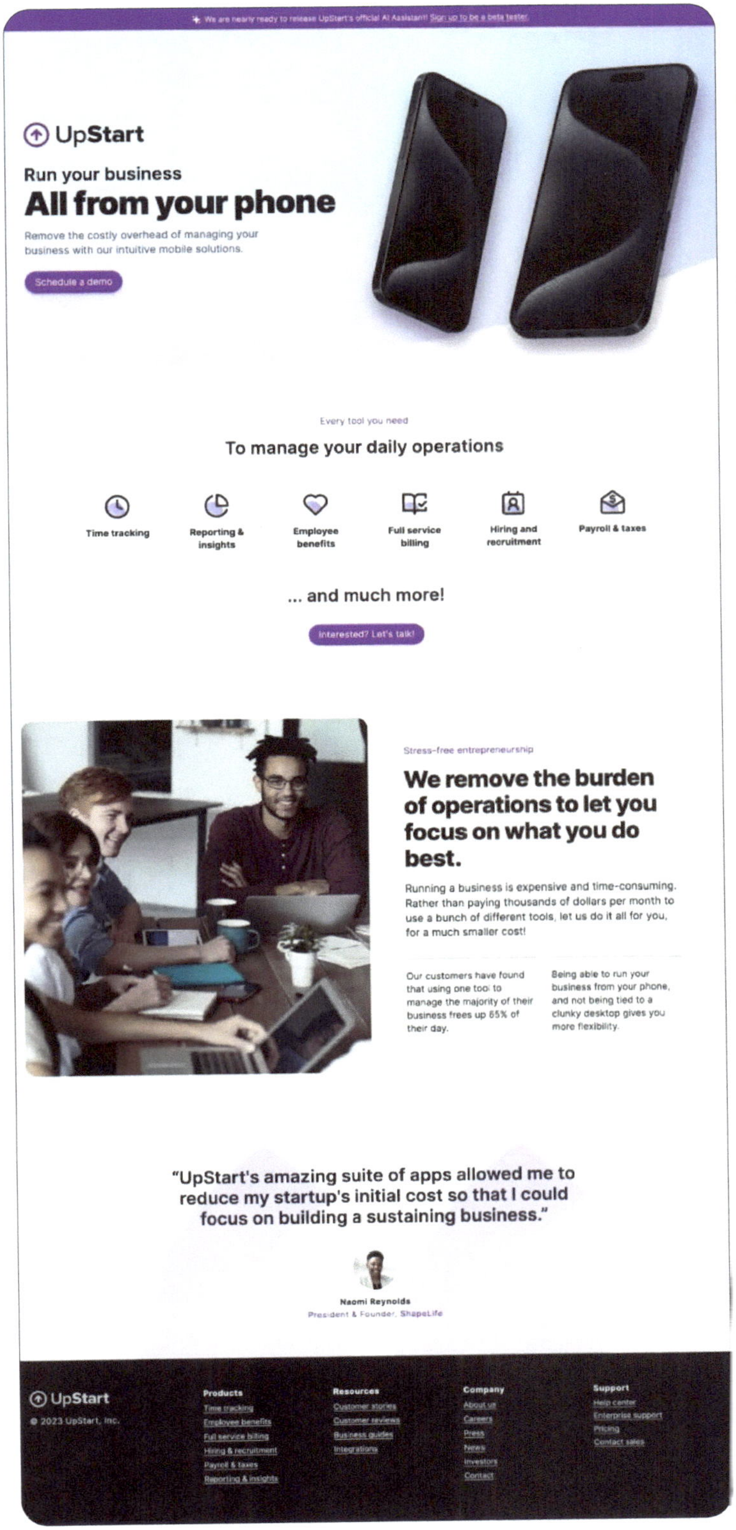

This site is an Experience Cloud site. See how pretty it can be?

Image courtesy a great Learn Experience Cloud article by Jason Kilp. Go to learnexperiencecloud.com and search for "Design Beautiful Responsive Web Pages with Lightning Web Runtime."

BEWARE...

...the shadow DOM.

A designer's worst nightmare...

The shadow DOM. If it's not our worst nightmare, it's right up there with being asked to use Comic Sans or make the logo bigger.

You're going to run into issues if you want to drastically change the look or behavior of any component. Why? The stupid freakin' shadow DOM. I hate it with all of my heart.

Here's why. Behold, an SLDS badge:

Pickles

Normally, if I wanted to make "Pickles" all uppercase, the code would look something like this (if you code, treat "badge" as a "div"):

```
badge .pickle-text {
    text-transform: uppercase;
}
```

And then, voila, you would get PICKLES in the badge. What the shadow DOM does is basically put Harry Potter's invisibility cloak around the the code inside of "badge" and any regular CSS that a developer tries to use won't see the pickle text...and "Pickles" remains "Pickles".

You might be wondering why the hell the shadow DOM even exists. You're not alone. Developers everywhere wonder the same thing. Salesforce says it's to protect the code that's hidden behind the invisibility cloak from other code "messing it up" effectively removing "cascading" from Cascading Style Sheets. I say that whomever the idea came from, clearly, they're in Slytherin.

There IS a way around the shadow DOM in many cases: styling hooks. You can learn all about them here:

www.lightningdesignsystem.com/platforms/lightning/styling-hooks

If you have frond-end dev chops, check it out. If not, make sure the developers you work with know about this and use it.

If you're lucky, you have genius engineers who know how to custom build components in a way that skirts the shadow DOM entirely.

Get to know the Experience Builder. Seriously.

The best way to design scalable UX for Experience sites is to set up your own and muck about in it. Let's get you setup in your own dev org.

One of the cool things about Salesforce is that every human can set up their own fully-functional development org for free. It even has sample data in it. If you want to tinker as you go through this, head over here and get your own:

developer.salesforce.com/signup

That page will say that someone in the Salesforce team "will be in touch to help you," but don't worry too much about that. I haven't had anyone contact me yet and I'm thankful for it. Once you go through that process and have logged in successfully, you can follow along.

This is really the only section where I'll give you step-by-step instructions because Experience Cloud is intensely different from all of the rest of the Salesforce ecosystem. And chances are that this is where you'll start your Salesforce Design life. So, let's get into it.

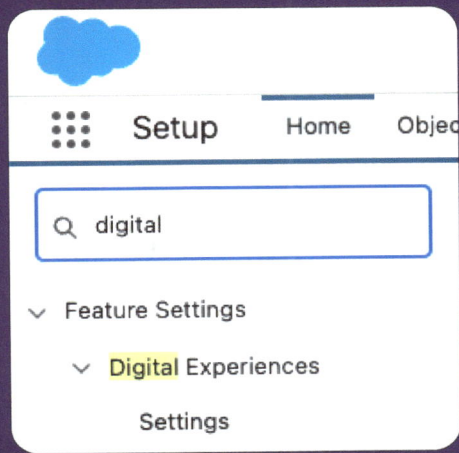

Turn on Digital Experiences.

That's right. They aren't on by default. Weird.

In the left sidebar there's a search field.

- Type in "digital"
- Choose "Settings" under Digital Experiences

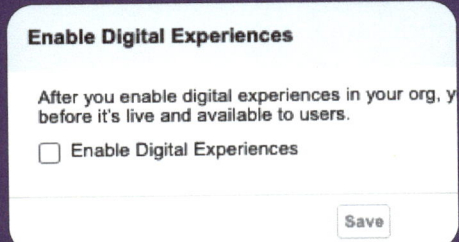

- Check the box
- Hit "Save"

Create a "digital experience."

Hit "New" and choose whatever template you want to start with. You can change it later and you can have more than one site. (You can have 100 of them.)

The hidden gotcha of Experience Cloud templates...

In my career, I've also designed and developed WordPress sites, so I'm used to themes and templates. But Salesforce does them a little weird. (Are you surprised?)

I mean, in theory, the way templates and themes work in Experience Cloud are kinda standard. Just remember that *these* themes are different from the *org themes* I told you about on page 36. So when you're talking to another person about themes, you might have to specify **which** theme you're talking about: org or Experience Cloud.

Because who needs a shared language anyway?

And with templates? Well, not all components are available in all templates out-of-the-box. Each template gives you access to a different set. Want to use the enhanced list views in the Customer Service template? It's not available there, but you **do** get access enhanced list views by default in Partner Central. If you want the functionality that component has in Customer Service, you'll need a custom component built.

Why am I telling you this?

It all comes back to: does the client have the time and money to build that beautiful site you just prototyped?

It's worth getting hands-on in the experience builder and experimenting with Experience sites in general.

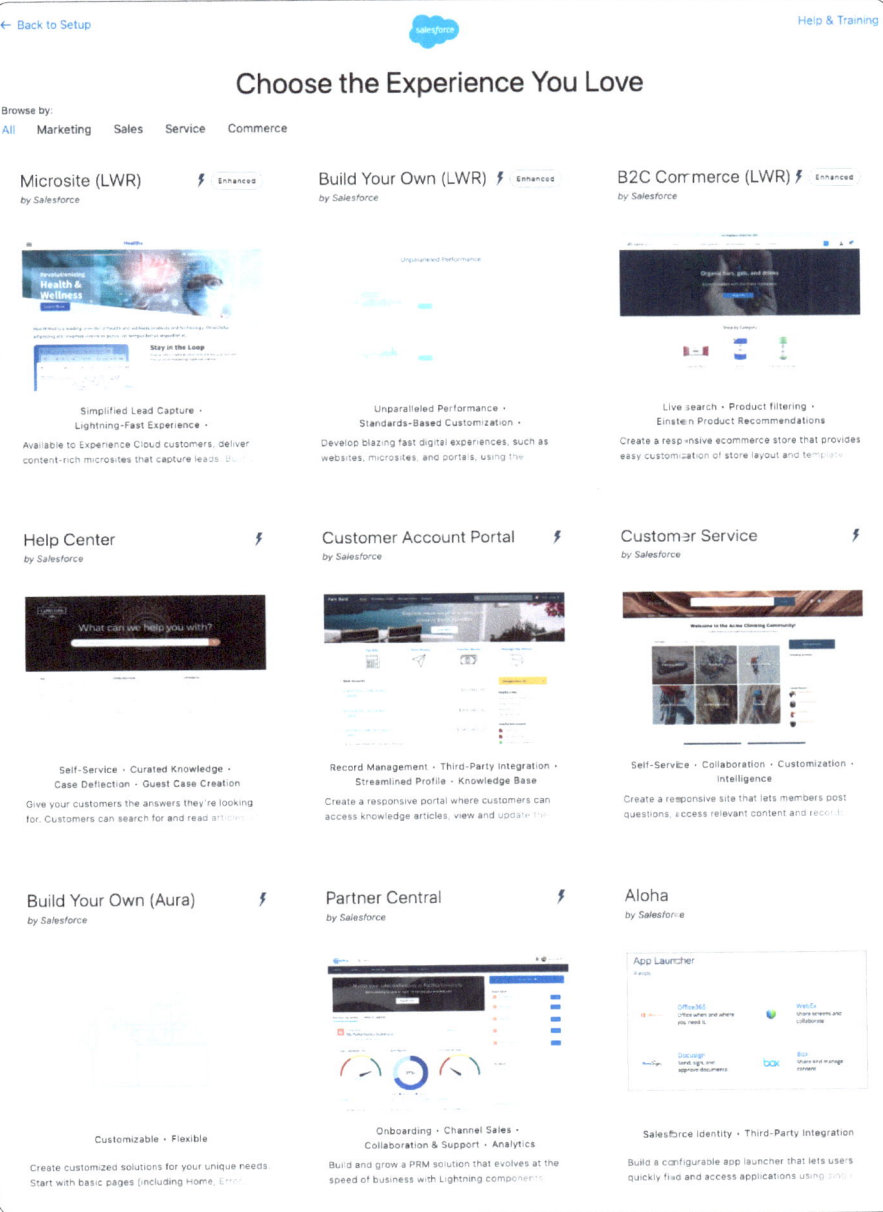

These are the templates you can choose from and many are self-explanatory.

But you'll see two Build Your Own options: Aura and LWR. The way the code is structured is fundamentally different. Aura has been around for a LONG time and, when possible, we stay away from it.

LWR stands for Lightning Web Runtime. You don't need to remember that. It's newer, faster, and more secure and more accessible than Aura. If the template doesn't have (LWR) in the name, it runs on Aura.

Getting around the builder

You made it! You're in the builder now, right? No? No worries, you can always reread this later. I'm happy to keep asking you the same question every time you come to this page.

When you first pop into the builder, it might throw some popup errors at you. That's ok. Ignore them, they don't matter.

After you've closed out the annoying modals, you're left with <snark> the beautiful and extremely usable builder interface. </snark> It's got a few quirks, those error messages being one of them. I'll get into the others shortly. For now...welcome.

The top bar

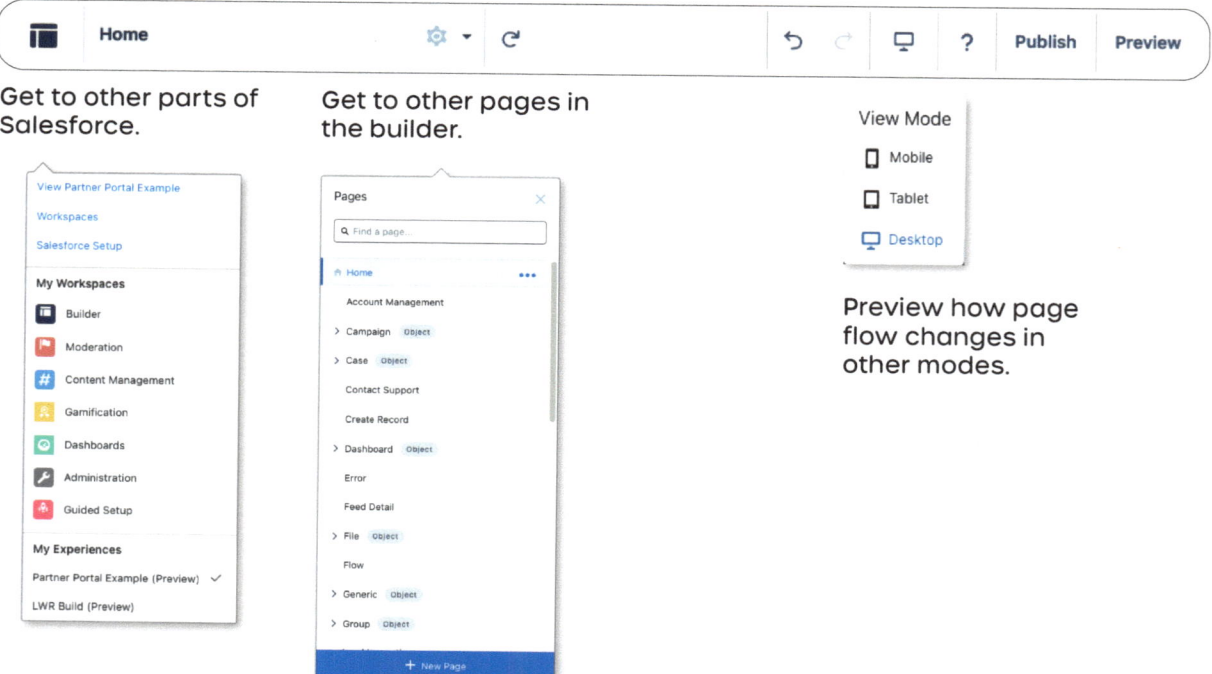

Get to other parts of Salesforce.

Get to other pages in the builder.

Preview how page flow changes in other modes.

The floating side menu.

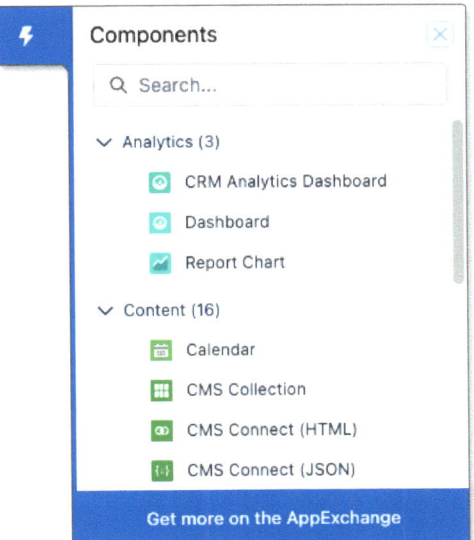

The components pane shows you what widgets you can add to the page. Just drag and drop, but sometimes you have to drag and drop twice for the builder to pay attention.

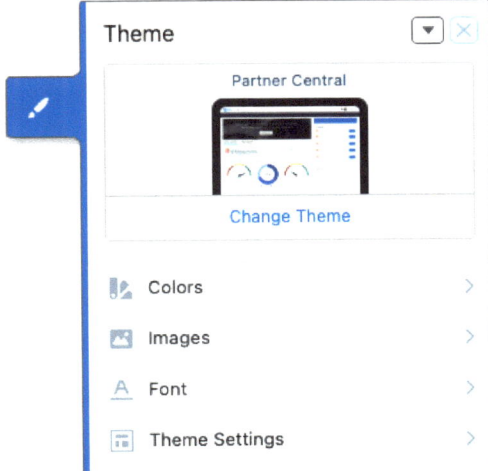

The theme pane has a lot of options to update the look of your experience site overall. I'll get into this more on the next page.

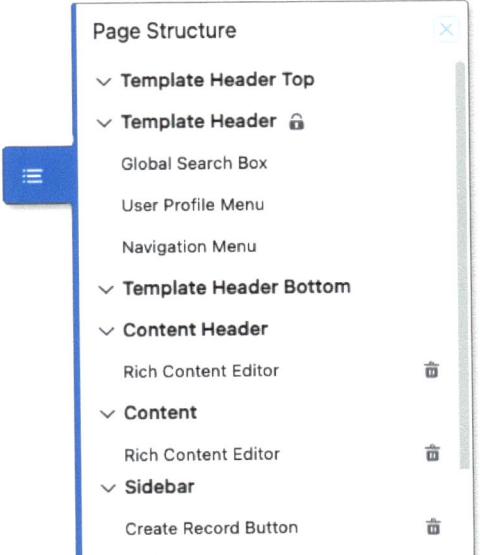

The page structure panel will show you what components are in each section of the page you're on. It's especially handy when there's a glitch and you can't select the component on the page itself.

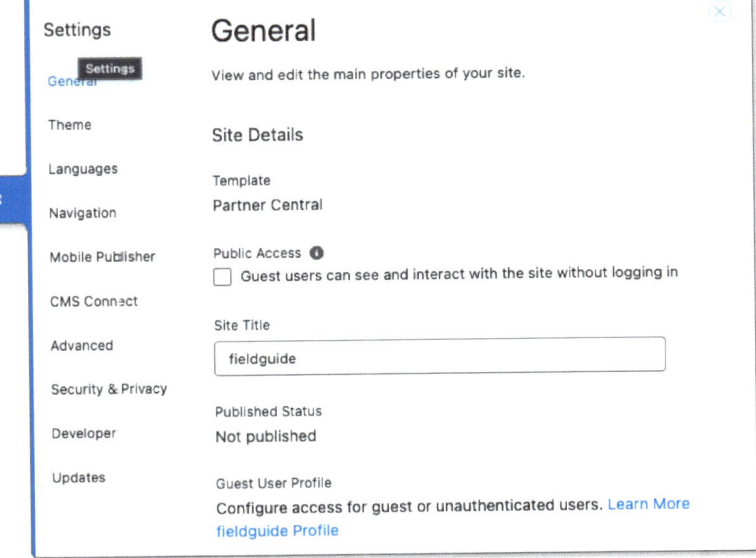

General settings has all of the most boring things stuck together. Noodle around in there to check it out. You can change **themes** here, odd that you don't get that option when you're first building the site.

Experience Site Themes

Themes within templates. Each template has a default theme. You won't get to choose the template theme when you're setting up the site, but you can alter the default one here OR choose a different one entirely.

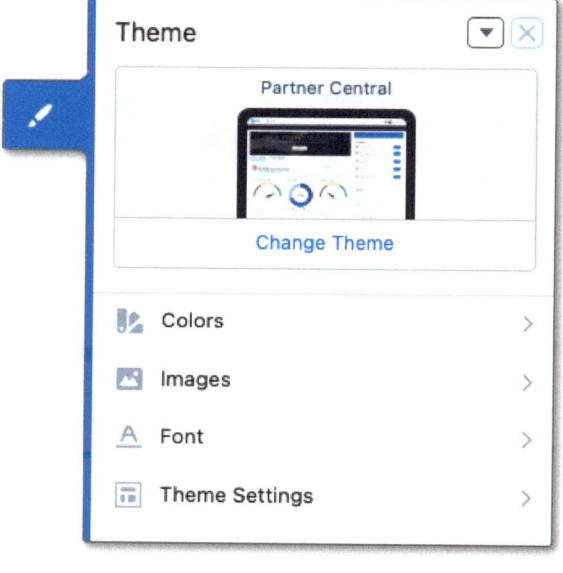

The first thing you'll see and where you can change the overall theme if you want. You'll typically have 7 to choose from. You can save out custom themes.

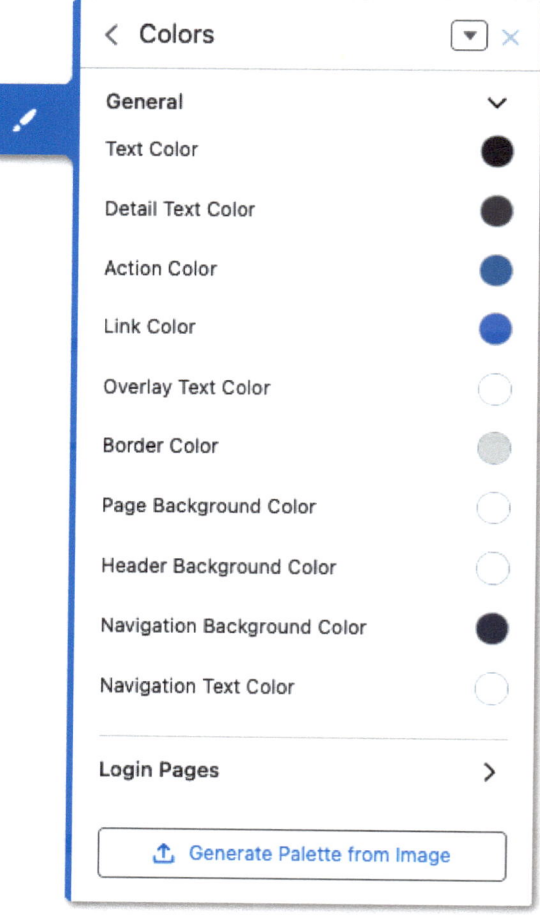

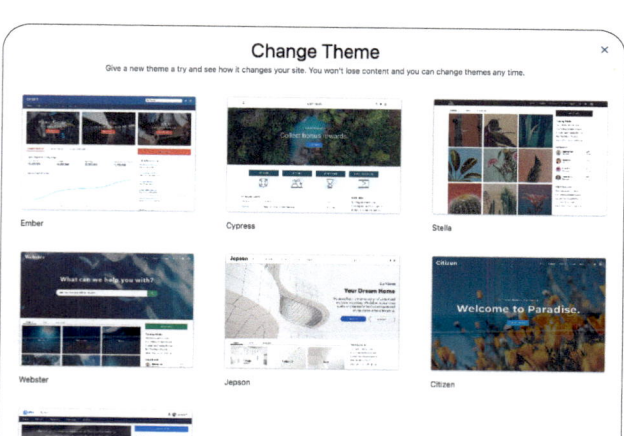

There are 10 colors you can set in the theme which will apply to the whole site. "Action color" is going to act as a primary color. Any other colors will have to be coded in.

(If your site is all LWR, this pane gives you WAY more options.)

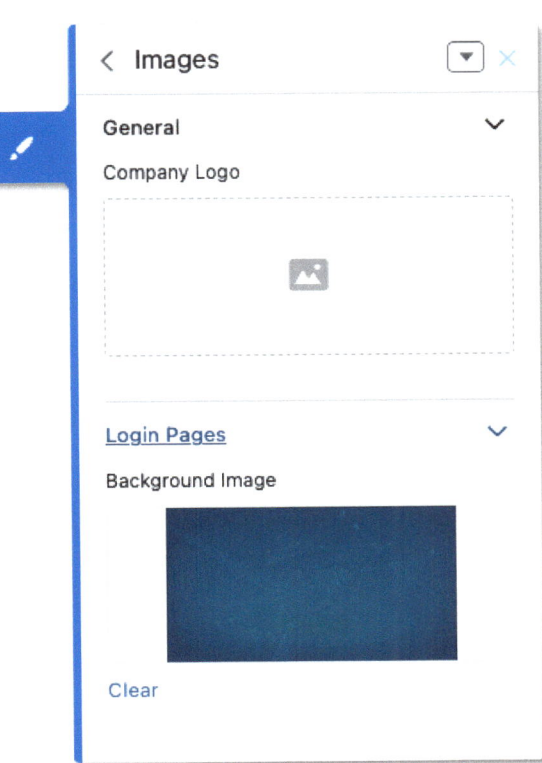

Look! You can set a background image! Ew. But hey, the logo is useful.

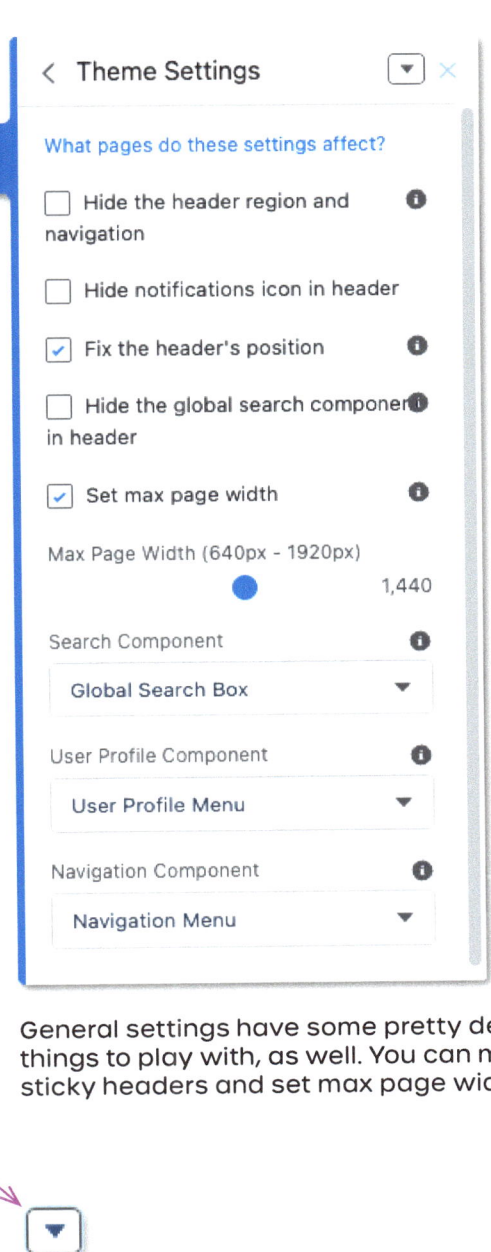

General settings have some pretty decent things to play with, as well. You can make sticky headers and set max page widths.

Manage Branding Sets

</> Edit CSS

Branding sets allow you to show different theme settings depending on the user. You can add custom CSS. It shouldn't live here permanently, but it's a good place to play.

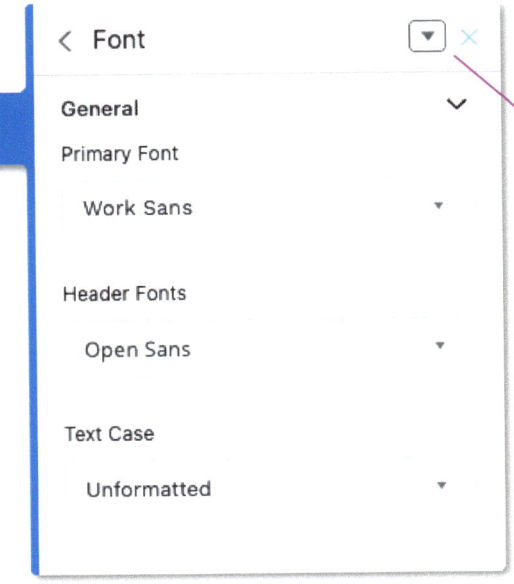

You can choose from a list of 44 fonts including *Comic Sans*.

You can configure components easily... to a point.

One thing I DO like about page components is that most, if not all, have options that you can change/select without any code just by clicking on the component.

Your'e even likely to see this in any custom components your engineers have developed or installed from the Salesforce AppExchange.

One thing that does annoy me from a pure UX POV is that the configuration panes for standard components are organized nicely in ways that the pane doesn't grow so long that you're scrolling for a long time to get to the option you want. Custom components are just a list and can scroll for a bit.

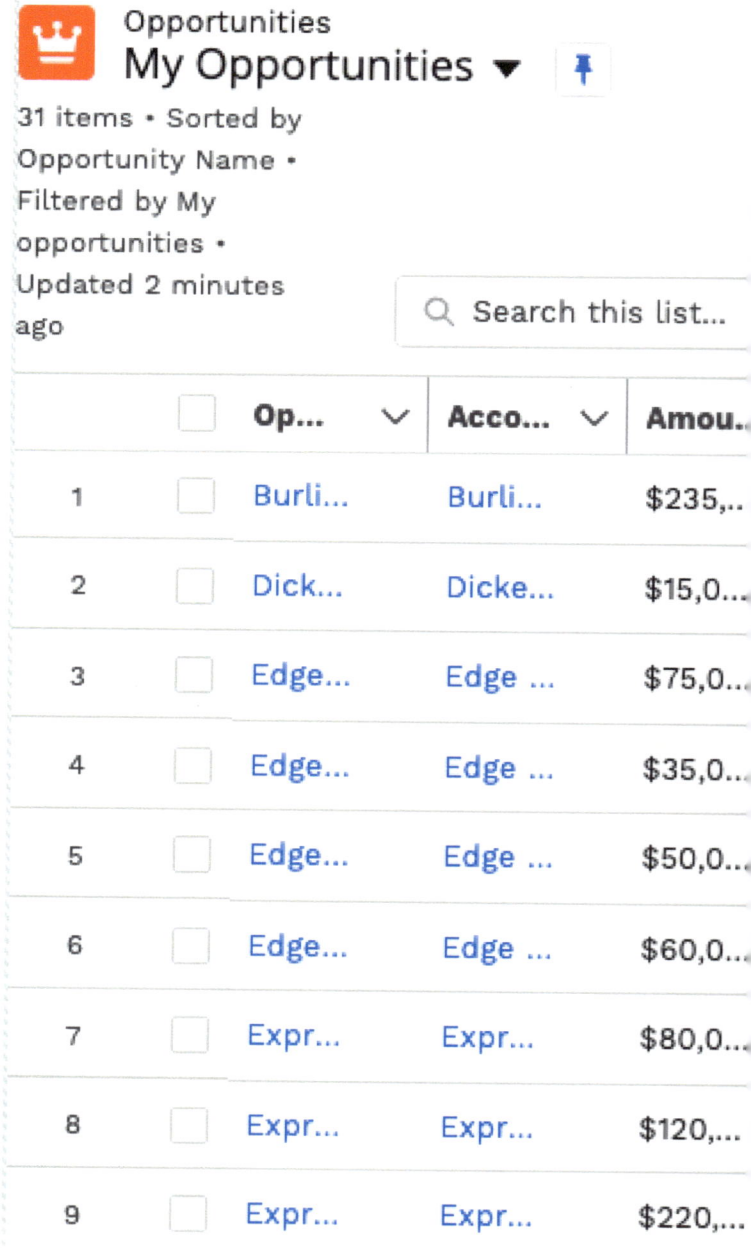

Let's take this standard list view as an example. There are a bunch of features you can turn on or off, determine how complex or simple the table looks, and add or remove action buttons.

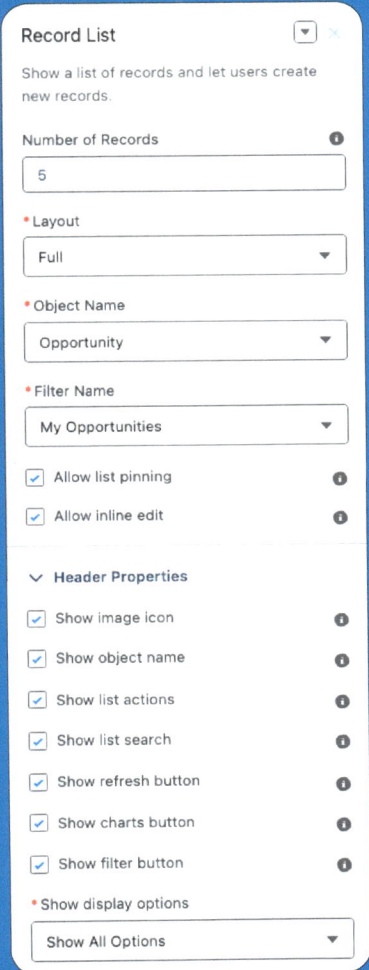

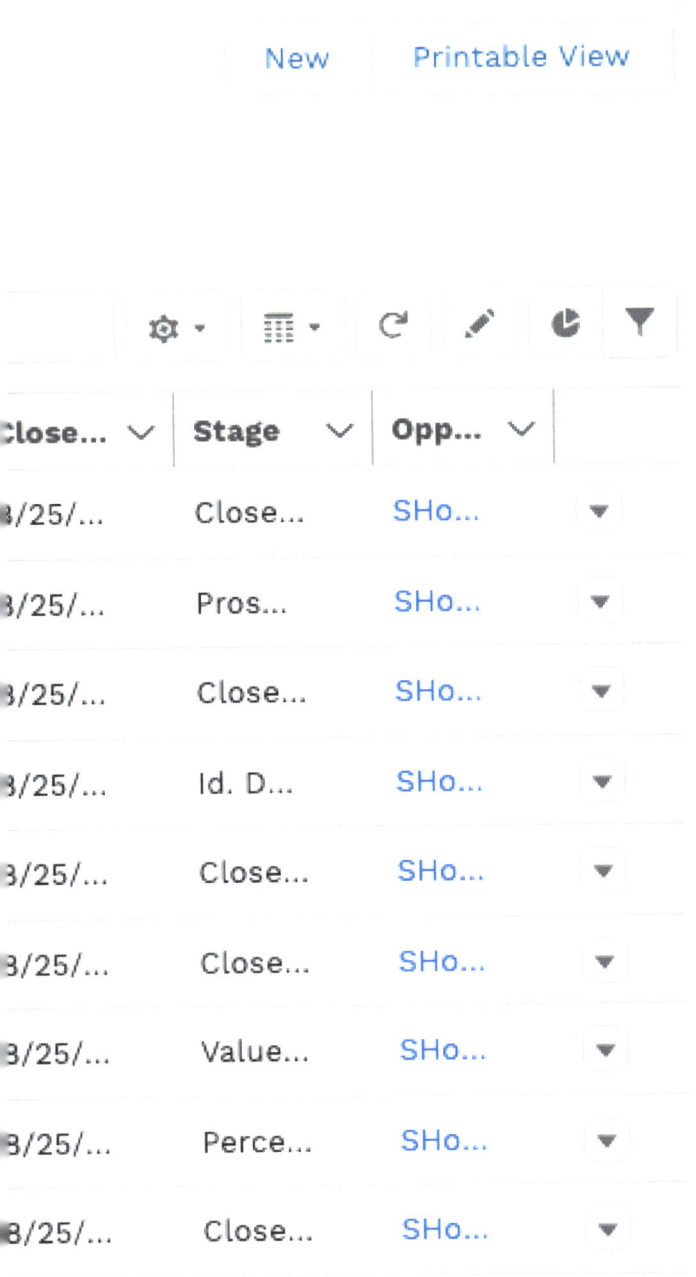

This is the configuration pane for this list. You can tell it's a default component from Salesforce because you can collapse the Header Properties.

You're an interpreter. A peacekeeper. ...a hero?

The users have needs. The business has goals. Quality Assurance wants everything functional and to make sense. The engineers? They can be your greatest asset or the biggest pain in your asset. And it's your job to get them all to understand that the users come first.

You're not just designing; you're bridging worlds.

Your mission, should you choose to accept it, is to navigate the waters between user needs, stakeholder dreams, and the "just let me code" desires of engineers. Welcome to the Bermuda Triangle of design.

This is the case in any project you're on, but especially important when they're Salesforce projects. **You** know the user needs and the business goals. But you also have to deal with stakeholders' hopes, dreams, and demands.

You'll hear it a lot: We **must** have feature ABC in the release, whether it's in the MVP or not. Then scope creeps, priorities change, and suddenly, the engineers have to spend a couple of months building and testing a custom connector between Salesforce and some propriatary software.

Somehow, you need to communicate the crawl, walk, run paradigm to the stakeholder and convince them to crawl first. But how?

You'll like this one: www.fastgood.cheap

Champagne Dreams on a Beer Budget

Your client's eyes sparkle with visions of grandeur, yet the budget whispers, "Cup o' Noodles for dinner."

Tools of the Trade:
- Use the "Good, Fast, Cheap: Pick Two" triangle to explain project constraints.
- Introduce the concept of "phased implementation" with "let's build the foundation before we build the roof."

The "MVP is ALL THE THINGS"

Your client dreams of an MVP that's essentially impossible—embodying every feature imaginable, yet somehow fitting into two sprints and a tight budget.

Tools of the Trade:
- Prioritization workshops will help out, but you have to go into it understanding Salesforce's limitations
- Make sure you're armed with a "Level of Effort vs. Impact" chart. Then remind the stakeholders "impact" takes both The Business and The User into account. Because happy users are good for business.

Engineers' Code Fever

Engineers love to code. Give them a problem, and they'll code a solution before you can say "declarative configuration." But here, in Salesforce land, not everything needs (or should) use custom code. You can still have custom solutions for complex problems, you'll just be working with screen flows, dynamic forms, and Lightning pages instead.

Tools of the Trade:
- Offer a peace treaty: a workshop that demystifies the power of Salesforce's declarative capabilities.
- And then determine what really needs to be custom code.

The "Simple" Illusion

"What do you mean, it's complex?" asks the client, thinking their request is as simple as flipping a switch. "I just need you to design one page for a form" sounds like a simple ask. You and I know that, like potato chip and tattoos, it's NEVER just one. A form page might have error states, modals, and conditional modals are that also need to be mocked up and inevitably you end up with 10 screens.

Tools of the Trade:
- A playful diagram showing what "simple" changes actually entail.
- Seriously, team up with engineering and then present your solutions to the client in a way where they can't say no.

The show must go on after handoff.

Ah, the handoff. That ceremonious moment when UX designers, with a mix of pride and relief, pass the baton to the engineering team. You might think it's time to kick back, pop the champagne, and watch your beautiful designs come to life. Well, pour that bubbly back into the bottle, friends. The reality? Our show is far from over.

The "Close Enough" Dilemma

In a world where stakeholders clutch their wallets tighter than their morning coffee, the first on the chopping block in any project are often those unsung heroes in Quality Assurance. But little do our stakeholders realize, QA is the guardian of user experience, the final checkpoint before our creations venture into the wild.

Am I being overly dramatic? Yes. Yes I am. But only in HOW I'm saying it and not what I'm saying. Please always advocate for QA to stay.

Yadda, Yadda, Yadda... Collaboration

It's not a one-off meeting; it's a continuous cycle of feedback and iteration. Team up with your QA team during the build and test phases.

- **Early Engagement:** Jump into the QA process from the start.

- **Share the Prototypes:** Hand over your design specs, interactive prototypes, and any secret handshakes that help QA navigate the world you've crafted. Your work is the treasure map; make sure QA has the latest version.

Keeping the UX Flame Alive

Our role is to be the guardians of usability from concept to launch, ensuring that the users will end up with a fantastic experience, right? So...fun tips I remind myself of often:

- **Accessibility is non-negotiable.** Just because something works doesn't mean it works for everyone. Be an accessible design champion. This is a hill I will always choose to die on.

- **Don't design for your ego.** Sometimes, what we envisioned doesn't translate perfectly into the real world and it's okay. Our egos might bruise, but usability always take the front seat. Remember, we're designing for users, not design awards.

And please remember, the handoff isn't a goodbye. It's more of a "See you in two minutes because we've got more work to do."

Congratulations.

You're now
part of the

0.5%

of the UX
industry who
can make
Salesforce less
Salesforcey.

About the author

I don't like writing about myself. So instead, I fed ChatGPT a bunch of feedback from talks I've given and a little bit about myself and let the AI write it. Here's what it had to say...

ChatGPT's self portrait.

In the digital realm, where the intersection of design and technology sparks innovation, Stephy shines as a beacon of creativity and inclusivity. Her journey from the structured world of chemistry to the fluid and dynamic universe of UX design is a testament to her versatility and insatiable curiosity. After more than two decades of navigating the intricacies of design, Stephy has become synonymous with solutions that are not just visually appealing but are profoundly impactful, making the digital space accessible to everyone.

Stephy's foray into the world of accessible digital design wasn't a chance occurrence but a response to the challenges faced by those closest to her—her children and coworkers grappling with the nuances of remote learning and digital accessibility. It's this personal connection to the cause that fuels her passion for creating digital experiences that aren't just functional but genuinely inclusive. Sparkles and glitter may adorn her designs, but at their core, they embody the principles of empathy and communication—traits that Stephy champions as the cornerstone of effective design.

Beyond her role as a designer, Stephy is a visionary speaker, captivating audiences with her deep insights into UX/UI trends, accessibility guidelines, and the

transformative power of thoughtful design. Her presentations, lauded for their engaging delivery and practical applications, serve as a bridge connecting the dots between design theory and real-world impact. *(Hmmm... that sounds like a great talk I should give.)* Whether it's improving PowerPoint presentations for enhanced accessibility or advocating for DEI within the design space, Stephy's contributions are both a source of inspiration and a call to action for her peers.

To witness Stephy in her element is to see a pro streamer and presenter in action. Feedback from her sessions underscores her knack for creating engaging and informative experiences, with attendees often leaving with a renewed enthusiasm for tackling the challenges of digital design and presentation.

But let's remember, underneath the pro-streamer facade and the ability to make PowerPoint bend to her will, she's just someone who figured out how to turn "shitty digital experiences" into digital gold. And she did it all while championing the cause of accessibility, because, as it turns out, making things easy for everyone to use is actually pretty cool.

Stephy's approach to design is unorthodox yet profoundly effective. A chemist turned designer, she sees beyond the surface, understanding that at the heart of every great design is the ability to connect, communicate, and empathize. Her mantra that "design degrees don't matter at all" underscores her belief in the universal language of design—a language shaped not by credentials but by the ability to listen, understand, and respond to the needs of the user.

In a world where digital experiences are increasingly central to our daily lives, Stephy stands out as a guardian of accessibility, a designer par excellence *(ok, ChatGPT is laying it on THICK)*, and a storyteller whose narratives are woven into the fabric of her designs. Her commitment to creating spaces that welcome everyone is not just her profession; it's her passion—a passion that sparkles as brightly as the designs she creates.

But seriously, can someone please tell me why
Salesforce's default body copy size is THIRTEEN pixels?!

$39.99

ISBN 979-8-9902657-0-7

53999>

9 798990 265707

www.ingramcontent.com/pod-product-compliance
Lightning Source LLC
Chambersburg PA
CBHW041455120626
46547CB00003B/447